Assessment of Data Quality Needs for use in Transportation Applications

By Hesham Rakha, Hao Chen, Ali Haghani and Kaveh Farokhi Sadabadi

Virginia Polytechnic Institute and State University

and

University of Maryland, College Park

1. Report No. MAUTC-2011-01	2. Government Accession No.	3. Recipient's Catalog No.
4. Title and Subtitle Assessment of Data Quality Needs for use in Transportation Applications		5. Report Date February 2013
		6. Performing Organization Code
7. Author(s) Hesham Rakha, Hao Chen, Ali Haghani and Kaveh Farokhi Sadabadi		8. Performing Organization Report No.
9. Performing Organization Name and Address Virginia Polytechnic Institute & State University Blacksburg, VA 24061 University of Maryland College Park, MD 20742		10. Work Unit No. (TRAIS)
		11. Contract or Grant No. DTRT07-G-0003
12. Sponsoring Agency Name and Address US Department of Transportation Research & Innovative Technology Admin UTC Program, RDT-30 1200 New Jersey Ave., SE Washington, DC 20590		13. Type of Report and Period Covered Final, - 6/1/2011 - 2/28/2013
		14. Sponsoring Agency Code

15. Supplementary Notes

16. Abstract
The objective of this project is to investigate the data quality measures and how they are applied to travel time prediction. This project showcases a short-term travel time prediction method that takes into account the data needs of the real-time applications. The objective of this research is to prepare and disseminate accurate short-term (up to 15 minutes ahead) travel time predictions on major highway corridors in the state of Maryland using real-time and archived Bluetooth travel time samples, probe-based INRIX data, and stationary sensor data pooled together in Regional Integrated Transportation Information System (RITIS). In addition the research effort also develops a medium-term travel time prediction algorithm using pattern recognition techniques. The algorithm is used to predict travel times between Richmond and Virginia Beach in the state of Virginia. Unlike previous studies that use travel time as the variable, the traffic state spatiotemporal evolution is used to predict traffic patterns. The approach uses traffic state data for the current day to match with a historical data set to identify similar traffic patterns and predict travel times into the future. The tasks of this study start from data collection and analysis. The raw INRIX data, including data from I-64 and I-264 between Richmond to Virginia Beach for the past three years, are used in this study. Several problems with the raw data are analyzed, including geographically inconsistent sections, irregular time intervals of data collection, and missing data. Subsequently, a travel database is constructed to obtain daily spatiotemporal traffic states in which traffic state information and dynamic travel times are included. A travel time prediction algorithm is developed using speed measurements and which fully utilizes the relationship between traffic state and travel time. INRIX data for the selected 37-mile freeway stretch (Newport News to Virginia Beach) are used to test the proposed algorithm. The testing results indicate that the proposed algorithm outperforms the other three methods including using instantaneous measurements, using a Kalman filter, and using the k-nearest-neighbor method. Moreover, the case study on the entire 95-mile freeway stretch from Richmond to Virginia Beach demonstrates the superiority of the proposed algorithm over the instantaneous approach that is currently used by VDOT. The proposed prediction method reduces the prediction error by approximately 50 percent compared to the current instantaneous method, especially at the shoulders of the peak periods.

17. Key Words Traffic state, travel time, traffic prediction, pattern recognition, state space, Kalman filter, NGSIM		18. Distribution Statement No restrictions. This document is available from the National Technical Information Service, Springfield, VA 22161	
19. Security Classif. (of this report) Unclassified	20. Security Classif. (of this page) Unclassified	21. No. of Pages 123	22. Price

Disclaimer

The contents of this report reflect the views of the authors, who are responsible for the facts and the accuracy of the information presented herein. This document is disseminated under the sponsorship of the U.S. Department of Transportation's University Transportation Centers Program, in the interest of information exchange. The U.S. Government assumes no liability for the contents or use thereof.

Table of Contents

1 Introduction 1
   1.1   Assessment of Traffic Data Quality 1
   1.2   Motivation for Dynamic Travel Time Prediction 2
2 Literature Review 2
   2.1   Models Based on Eulerian Data 4
   2.2   Models Based on Lagrangean Data 11
   2.3   Models Based on Integrated Lagrangean Data 11
   2.4   Models Based on Eulerian and Integrated Lagrangean Data 15
   2.5   State Space Models 20
   2.6   Bayesian Filter for Traffic State Estimation and Prediction 25
   2.7   Dynamic Travel Time Prediction 27
3 Methodology of Traffic Estimation 30
   3.1   Traffic Model (LWR-v) 31
   3.2   Integrated Lagrangean (Travel Time) Data Representation 36
   3.3   State Space Models 43
   3.4   Optimal Estimation Methods 47
   3.5   Summary 56
4 Methodology of Dynamic Travel Time Prediction 56
   4.1   The Dynamic Travel Time Prediction Framework 57
   4.2   Matching Traffic Patterns 57
   4.3   Dynamic Travel Time Prediction 58
   4.4   Revised Algorithm 61
5 Findings of Traffic Estimation 62
   5.1   Traffic Datasets 62
   5.2   Traffic Modeling 63
   5.3   Traffic Data Fusion 64
   5.4   Travel Time Model and Finite Difference Schemes 66
   5.5   Summary 72
6 Findings of Dynamic Travel Time Prediction 72
   6.1   Data Collection and Analysis 72
   6.2   Travel Database Construction 79
   6.3   Test Environment 85
   6.4   Case Study 1 85
   6.5   Case Study 2 90
   6.6   Case Study 3 98
7 Conclusions 104
   7.1   Traffic Estimation 104
   7.2   Dynamic Travel Time Prediction 105
8 References 107

List of Figures

Figure 1. Cumulative input-output curves concept. 4
Figure 2. Illustration of typical speed-based travel time prediction concepts. 8
Figure 3. Greenshield's speed-flow-density relationships. 32
Figure 4. Concept of travel time as distance from downstream boundary in a wave propagation paradigm. 37
Figure 5. Schematic illustration of a vehicle trajectory in space-time domain 39
Figure 6. Space-time grid representation of the solution domain. 42
Figure 7. Framework of proposed dynamic travel time prediction algorithm. 57
Figure 8. Illustration of dynamic travel time. 59
Figure 9. The flow chart of dynamic travel time calculation. 60
Figure 10. Samples of current traffic status and selected candidates......................................61
Figure 11. Mean absolute errors of the estimated speeds at elemental nodes of US-101 (blue and red lines represent errors from FDM and FEM methods, respectively) 64
Figure 12. DKF speed estimate qualities at different penetration rates. 65
Figure 13. H_∞ speed estimate qualities at different penetration rates. 66
Figure 14. Spatiotemporal and boundary observations on US-101 mainlines. (upstream and downstream values are represented by blue and red lines, respectively)................................67
Figure 15. Travel time estimates based on observed speeds on US-101 mainlines (top: spatiotemporal travel time estimates, bottom: mean absolute error) 69
Figure 16. Travel time estimates based on estimated speeds on US-101 mainlines (top: spatiotemporal travel time estimates, bottom: mean absolute error) 71
Figure 17. INRIX data for I-64 and I-264. 73
Figure 18. Aggregated INRIX data for I-64 from RITIS website. 74
Figure 19. Aggregated INRIX data for I-264 from RITIS website. 75
Figure 20. INRIX raw data. 76
Figure 21. Sorted freeway sections along I-64. 77
Figure 22. Geographically inconsistent sample sections. 78
Figure 23. Sample irregular time interval of raw data. 79
Figure 24. Freeway stretch for Travel Database 1. 80
Figure 25. Samples of daily spatiotemporal speed map for Travel Database 1. 81
Figure 26. Freeway stretch for Travel Database 2. 82
Figure 27. Samples of daily spatiotemporal speed map for Travel Database 2. 84
Figure 28. Selected 37-mile freeway stretch for case studies 1 and 2. 85
Figure 29. Comparison of prediction results for a typical weekday (August 2, 2010)...........89
Figure 30. Comparison of prediction results for a typical weekend (August 7, 2010)...........90
Figure 31. The ground truth travel time data and the comparison of four methods for Monday. ..93
Figure 32. The ground truth travel time data and the comparison of four methods for Tuesday. 94
Figure 33. The ground truth travel time data and the comparison of four methods for Wednesday. 95
Figure 34. The ground truth travel time data and the comparison of four methods for Thursday. 95
Figure 35. Method error relative to ground truth for Friday. 96

Figure 36. Method error relative to ground truth for Saturday. 97
Figure 37. Method error relative to ground truth for Sunday. 97
Figure 38. Freeway stretch from Richmond to Virginia Beach along I-64 and I-264. 98
Figure 39. Spatiotemporal traffic state map and trip trajectories. 99
Figure 40. Average MAE of congested periods by two methods. 100
Figure 41. Maximum MAE by two methods for August 2010. 101
Figure 42. Travel time prediction results on August 04, 2010. 102
Figure 43. Predicted travel time distribution on August 04, 2010. 103
Figure 44. Travel time prediction results on August 27, 2010. 103

List of Tables

Table 1. Summary features of traffic speed/travel time estimation studies using state-space models. 24
Table 2. Overall performance of the approximate solution methods. 63
Table 3. Speed estimation quality 67
Table 4. Travel time estimation quality using observed speeds 68
Table 5. Travel time estimation quality using estimated speeds 70
Table 6. Relative errors by proposed travel time prediction method. 87
Table 7. Absolute errors by proposed travel time prediction method. 87
Table 8. Relative errors by KNN method. 88
Table 9. Absolute errors by KNN method. 88
Table 10. Prediction results of four methods. 91
Table 11. Prediction results of four methods for different time periods. 92

1 Introduction

1.1 Assessment of Traffic Data Quality

Traffic data collection, within the context of transportation operation and management, is becoming an increasingly valuable asset in today's transportation arena. Significant traffic data have been generated from Intelligent Transportation System (ITS) technologies in recent years and more data are anticipated through the emerging Connected Vehicle technology. The data have been widely utilized in managing system operations and providing information on traffic conditions. However, public and private users are finding that the utilization and operation of the data is an increasingly difficult task since the data are collected with different levels of accuracy and resolution, and data formats are incompatible. Furthermore, the problem worsens as the amount of data continues to grow. The quality of data in data collection, operation, and management efforts has resulted in the underutilization of data and increased utilization costs. Various problems were identified in recent research efforts regarding the quality of data for transportation operations, planning, traffic congestion information, transit and emergency vehicle management, and/or commercial truck operations.

Data quality has been questioned since the earliest stages of traffic data collection. Since a variety of ITS applications and various travel information systems have unique data requirements, the matter of data quality has become more urgent. Furthermore, in the last few years, this intricacy has been made more complex due the emergence of private services which are providing traffic information services to the public. Turner (2004) gave a definition of data quality as "the fitness of data for all purposes that required it. Measuring data quality requires an understanding of all intended purposes for that data". Traffic data have different meaning(s) to different consumers and the intended uses of data should be considered and understood when designing, implementing and operating data collection systems and applications.

There are numerous reasons for such deficiencies in provision of accurate and timely traffic data to general public. For instance, in the case of travel time data, apart from sensor errors, communication line failures, and use of naïve estimation methods, the fact that posted travel times are typically based on instantaneous realized travel times (as opposed to predictions) should be counted as the main contributor to the distrust among users and agency officials alike.

Traditional data collection systems may not assure the quality of data that satisfy the state-of-the-art transportation applications. There are urgent needs that the specific data quality measures should be considered for each traffic data application. This project will investigate the data quality measures for transportation data and present an overview of the requirements for the use of these data in various real-time and non-real-time transportation applications. As a case in point, this project specifically focuses on travel time data quality requirements. To ensure the highest quality travel time data, we propose that traffic data from different sources should be considered and every attempt should be made to complement stream of data from point sensors with more accurate data that might become available every once in a while. In this context, accuracy, update rate and intrinsic value of different traffic data sources should be considered.

1.2 Motivation for Dynamic Travel Time Prediction

Congestion has proven to be a serious problem across urban areas in the United States. In 2007, it cost highway users 4.2 billion extra hours of sitting in traffic and an extra 2.8 billion gallons of fuel. This all translated into an additional $87.2 billion in congestion costs for road users in 2007, which showed a 50% increase in cost compared to data from the previous decade. Even though the recent economic downturn is said to have marginally eased the congestion problem nationwide, new evidence shows an uptrend of traffic and, consequently, congestion is back (Schrank and Lomax 2007).

Tackling congestion (both recurrent and non-recurrent) has proven to be a challenge for highway agencies. Adding capacity in response to congestion is becoming less of an option for these agencies due to a combination of financial, environmental, and social issues. Therefore, the main focus has been on improving the performance of existing facilities through continuous monitoring and dissemination of traffic information. The minimum that can be accomplished is to inform the public or, specifically, the potential users of what they should expect on the roadways before and during their trips. Additionally, this information can be applied to provide alternatives to users so that they may make informed decisions about their trips. This is the essence of Advanced Traveler Information System (ATIS) applications such as 511 that have been implemented nationwide. In many states relevant traffic information is also posted on variable message signs (VMS) that are strategically positioned along the highways. However, the effectiveness of such efforts and the accuracy and usefulness of information they provide have been questioned (Peeta and J. L. Ramos 2006).

There are numerous reasons for such deficiencies in providing accurate and timely traffic data to the general public. Apart from sensor errors, communication line failures, and use of naïve estimation methods, the fact that posted travel times are typically based on instantaneous realized travel times (as opposed to predictions) should be counted as the main contributor to the distrust among users and agency officials alike. To remedy this situation, we propose that traffic data from different sources be considered and every attempt be made to complement stream data from point sensors with more accurate data that might become available at certain points in time every once in a while. In this context, the accuracy, update rate, and intrinsic value of different traffic data sources should be considered. Low-cost stationary traffic sensors are most prevalent but have greater measurement errors. Conversely, probes are essentially able to provide an accurate trajectory (time and location) of the vehicle as it passes through a road segment.

In addition to traffic estimation, this study uses privately available data from INRIX to predict travel times. Since the study area extends over 95 freeway miles the travel times are in the range of 2 hours during peak periods. This long prediction horizon makes predictions based on macroscopic modeling approaches inadequate. Consequently, historical data are used to develop a data-driven approach for mid-range predictions (e.g., 1- to 2-hour predictions).

2 Literature Review

Travel time estimation and prediction problem can be classified in many different ways:

First, they can be grouped based on the facility type on which the problem needs to be solved. For instance, travel time estimation in facilities with interrupted flow should be treated differently from estimations performed on facilities with uninterrupted flows. While a lot of effort in the past has been spent to estimate travel times in the latter facility types (e.g. freeways), very few studies report on methods to estimate or predict travel times over the interrupted facilities (e.g. arterials).

Second, in many practical cases, proposed methods are limited to data readily available from existing traffic sensing technologies. This would include stationary sources such as inductive loop detectors and road side microwave radars. Vehicle re-identification data from license-plate or toll-tag readers can provide a sample of travel times. Finally, probes are capable of not only providing a travel time sample but they also will give insight to the evolutions of traffic conditions over space and time inside the segment under study. Methods to fuse data from different sources and to establish a hybrid estimate of travel time are gaining more popularity.

The third aspect that can be used to distinguish between different travel time estimation and prediction methodologies is the inductive (non-parametric) or deductive (parametric) nature of the proposed methods. In broad terms, inductive methods are data-driven and make extensive use of historic observations. Given a representative data set, inductive methods are shown to have a good performance in predicting travel times under recurrent traffic conditions. On the other hand, deductive methods take into account physical principles governing traffic operations and resulting interactions between different traffic parameters and various external factors affecting traffic. Therefore, deductive methods are capable to handle unforeseen traffic situations and are equally useful in traffic control applications due to their normative nature as opposed to inductive models which have mere descriptive powers.

The fourth characteristic of reported models can be defined with regard to their adaptive or non-adaptive nature. In general, adaptive methods have more flexibility and are able to discern temporal changes in the traffic system under both recurrent and non-recurrent conditions. This is a very desirable feature since accurate travel time estimates are most needed when unforeseen conditions due to incidents, construction, inclement weather, and such arise.

Last, but not least, property of travel time models is the inclusion of a sound vehicular traffic model in the estimation process. Unfortunately, the majority of methods reported in the literature are solely based on generic statistical techniques and do not make any effort to take advantage of the existing knowledge on traffic flow theory.

Travel time estimation and prediction methods reported in the literature can be broadly classified into four groups according to their adopted methodology.

1. Conservation of flow
2. Kinematics
3. Statistical
4. Hybrid

2.1 Models Based on Eulerian Data

2.1.1 Conservation of Flow (Input-Output Curves)

The first group of methods for travel time estimation is based on the conservation of flow principle. Generally speaking, this principle states that vehicles entering a segment at upstream over some time along with the ones initially existing inside the segment are the ones that will leave the segment at the downstream during the same time or will remain in it at the end of the time period. This gives rise to the idea of obtaining travel times by comparing N-curves representing cumulative number of vehicles passing upstream (entering) and downstream (exiting) of the segment. This idea was first presented by Newell (1993) in which cumulative number of vehicle arrivals at a sequence of locations on a highway are used to estimate travel times, flow variations and shockwave creation and propagation. Cassidy and Windover (1995) describe a similar method for assessing the dynamics of freeway traffic. The methodology is more descriptive rather than normative (prescriptive). Figure 1 further illustrates the concept. In this figure, slope of the cumulative curves is equal to traffic flow (q); the vertical distance between two curves at each time represents the accumulation of vehicles on the segment (a) while the horizontal distance is equal to travel time (T) on the segment under study.

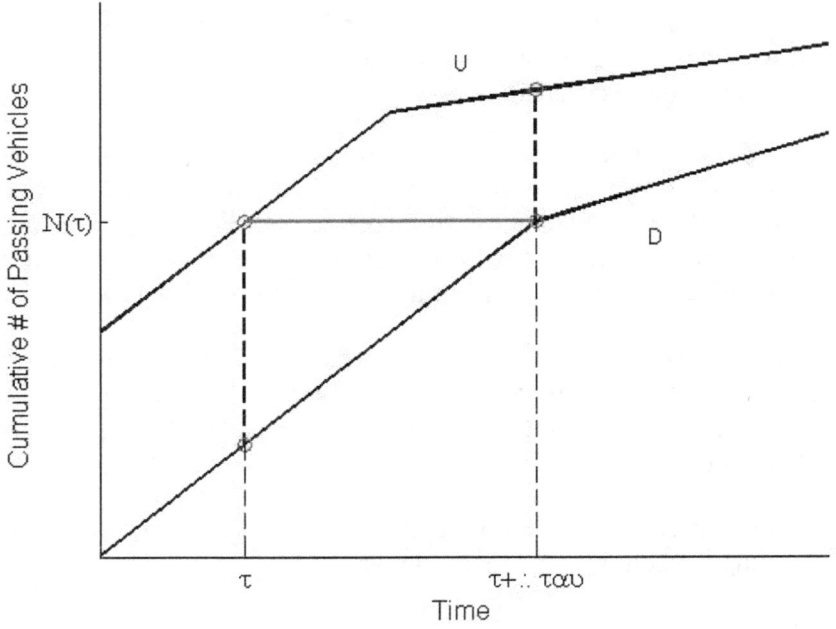

Figure 1. Cumulative input-output curves concept.

$$q(x,t) = \dot{N}(x,t) \tag{1}$$

$$N(U,t) = N(D, t + T(t)) \tag{2}$$

$$T(t) = N^{-1}(D, N(U,t)) - t \tag{3}$$

It may seem that counting the number of vehicles passing a point of the highway should be an easy process. Inductive loop detectors and a variety of stationary sensors are used to accomplish this task, but it is well-known that counts obtained using these technologies are less than perfect. In fact, ILDs which are not calibrated properly are susceptible to a phenomenon called drifting in which passage of some vehicles are missed. Such technological deficiencies along with the necessary knowledge of the initial number of vehicles in the segment for these methods to work have been the main impediments in widespread use of these methods. To this we can add the fact that the concept is primarily suitable for segments with no access/egress points in the middle. Otherwise, number of vehicles entering and exiting in the middle of the segment should be taken into account.

Assuming that cumulative curves are continuous and smooth everywhere, Astraita (1996) took the derivative of both sides of equation (2) and derived the following relation between flow rates at upstream, downstream and travel time on the segment. It should be noticed that flow rates are easier to obtain and to work with than the cumulative number of vehicles.

$$q(D, t + T(t)) = \frac{q(U,t)}{1+T'(t)} \tag{4}$$

Carey et al. (2003) propose a dynamic link travel time model based on the assumption that travel time is a non-decreasing function of the average surrounding flow experienced by a vehicle while traveling along the segment. They approximated this average flow as a linear combination of flow at the entrance and at the exit points of the segment as experienced by the vehicle.

$$T(t) = f(\beta q(U,t) + (1-\beta)q(D, t+T(t))) \tag{5}$$

After substituting for downstream flow rate using equation (4), they obtained the following model.

$$T(t) = f(\beta q(U,t) + (1-\beta)\frac{q(U,t)}{1+T'(t)}) \tag{6}$$

And, after inverting and rearranging they obtained the following first-order ordinary differential equation.

$$T'(t) = -\frac{f^{-1}(T) - q(U,t)}{f^{-1}(T) - \beta q(U,t)} \tag{7}$$

Carey (2004) shows that this model has some desirable properties, such as causality, first in first out (FIFO) and similarity to the static model when flows are constant. Carey and Ge (2007) examine several discrete time approximation methods for numerical solution of their proposed model (5). These approximations are in fact simple forward and backward differencing methods that are widely used for solving differential equations with no closed form analytical solutions. They point out that simple approximate solutions may be violating FIFO property. Therefore, to keep the FIFO property in approximate solutions, regardless of the size of discrete time intervals, an alternate differencing method is suggested which applies the backward differencing method while moving forward in time. They concluded that this model can be equivalently solved as a simple optimization problem at each time interval. The optimization problem can be solved using simple line search algorithms such as golden section search.

Vanajakshi and Rillet (2006) proposed an adjustment algorithm based on generalized reduced gradient (GRG) method to fix problems associated with accuracy of inductive loop detector records. In essence, this method attempts to make smallest changes in the readings while maintaining the condition that cumulative flow at successive detector points should be smaller or equal to that amount at upstream points. Also, the constraint for allowing practically possible maximum number of vehicles on any road segment at any time is enforced under this methodology. These two conditions in fact hold up conservation of flow principle in the traffic stream. Vanajakshi et al. (2009) used these adjusted detector readings to improve on the travel time estimation method originally proposed by Nam and Drew (1996, 1998, and 1999). Vanajakshi et al. (2009) suggest that the congested flow model (Nam and Drew 1998) should be used throughout, and that density to be estimated based on a source other than cumulative flows. They use the relationship between occupancy and density to estimate the latter. They report between 9 to 16 percent error in travel time estimates on a segment between two detector stations in their test case. This error increases up to 20 percent on a 2 mile corridor.

Waller et al (2007) adopt an ARIMA(3,1,2) to forecast inflows to the freeway segment under study, then they use a meso-simulation technique called cell transmission model (CTM) to simulate propagation and movements of vehicles inside the segment. Later, based on cumulative flow curves at the segment endpoints they are able to forecast travel time. On a 3 mile freeway segment, they reported 10 to 23 percent RMSE on travel times predicted 5 minutes ahead using this method when compared with travel times obtained from VISSIM micro-simulation.

2.1.2 Approximate Kinematic Models

The second group is comprised of kinematic models. Kinematics is a branch of mechanics which deals with motion without regard to forces or energies that may be exerted on the objects under study. The basic notion of kinematics is that point speed of a vehicle at any given time is equal to the derivative of its trajectory at that time. Therefore, we can derive the relation between distance traveled, speed and travel time in an integral form,

$$\dot{X}(t) = v(X(t), t) \tag{8}$$

$$X(T) = X(0) + \int_0^T v(X(s), s) ds \tag{9}$$

where,

$X(t)$ is the vehicle position at time t, and

$v(X(t), t)$ is the vehicle speed at time t.

The integral in equation (9) is difficult to estimate since in most cases the speed profile of a vehicle during its trip is not known. Instead, it is common to approximate this integral with point speed measurements at multiple points along the segment over which travel time is to be estimated. Specifically, in highway applications, speeds at upstream and downstream of the segment are usually available.

$$X(T) \cong X(0) + \left[\frac{v(X(0),0)+v(X(T),T)}{2}\right]T \tag{10}$$

Therefore, travel time can be estimated as

$$T \cong \frac{2L}{v(X(0),0)+v(X(0)+L,T)} \tag{11}$$

where, L is distance traveled or length of the segment $[L = X(T) - X(0)]$.

Equation (11) essentially suggests an iterative method to estimate travel times which is called dynamic time slice method in the literature **Invalid source specified.**. A further approximation of this formula would result in what is called instantaneous method in which downstream speed at the time vehicle enters the segment is used,

$$T \cong \frac{2L}{v(X(0),0)+v(X(0)+L,0)} \tag{12}$$

Figure 2 illustrates the times and locations for which speeds are available and are being used to predict travel times versus what speeds should be used. Obviously, these approximations only work under stable traffic conditions when there is not much change in vehicle speeds over space and time.

Lindveld et al. (2000) employed the harmonic mean of speeds to substitute the integral in equation (9)

$$X(T) \cong X(0) + \left[\frac{2}{\frac{1}{v(X(0),0)}+\frac{1}{v(X(T),T)}}\right]T \tag{13}$$

which results in the following estimate of travel time

$$T \cong \frac{L}{2}\left\{\frac{1}{v(X(0),0)} + \frac{1}{v(X(T),T)}\right\} \tag{14}$$

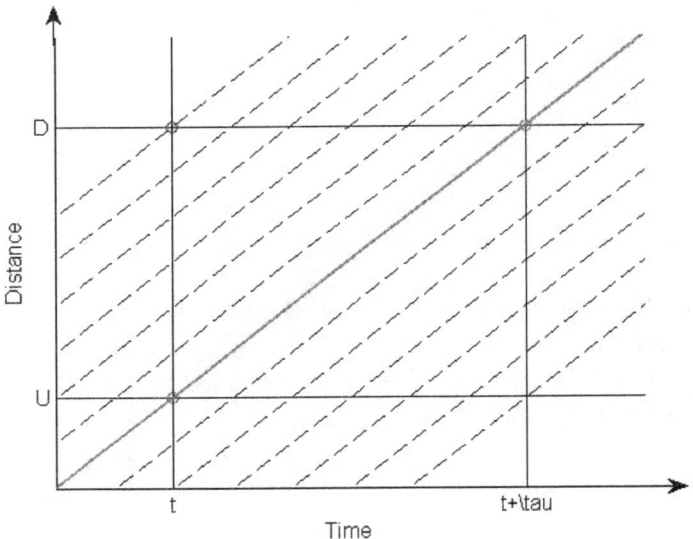

Figure 2. Illustration of typical speed-based travel time prediction concepts.

Further, they evaluated several kinematics based and flow correlation methods for travel time estimation and prediction in three different European sites (Amsterdam, Paris, and Padua-Venice). The input data for these methods generally comes from inductive loop detectors. Evaluation results show that these methods produce RMSEP in travel time estimation/prediction well above 10% under free flow conditions, while as congestion increases their performance rapidly deteriorates.

The kinematics methods are easy to use and provide inexpensive travel time estimation solutions which generally make use of existing sensing technologies and readily available data. However, they lose their accuracy as distances between consecutive sensing stations become large. Also, they are most accurate when traffic condition along the segment is stationary. As traffic conditions begin to change abruptly over time and/or space estimates from these methods become less reliable.

Various technologies are in use to measure vehicle speeds passing a given point on the highway. Inductive loop detectors are among the earliest sensors used for this purpose. In the single loop setting the relationship between detector occupancy, volume, and vehicle length can be used to estimate spot speeds. For a single vehicle, relationship between time it has kept the detector in presence mode (t_i), detector length (l_d), vehicle length (l_i) and its speed (v_i) is as follows,

$$t_i = \frac{l_d + l_i}{v_i} \tag{15}$$

However, one should keep in mind that data usually is not available at a single vehicle level; instead aggregate data (20~30 seconds) is typically provided by detectors. Therefore, occupancy of detector measured as fraction of time detector has been in presence mode in an interval is defined as below.

$$O = \frac{\sum_{i=1}^{q} t_i}{\Delta t} = \frac{1}{\Delta t}\sum_{i=1}^{q}\frac{l_d+l_i}{v_i} \qquad (16)$$

where, Δt is the time interval, and q represents the number of detected vehicles in that same interval.

Kurkjian et al. (1980) use an approach based on the first-order method of moments to estimate spot speeds using a single inductive loop detector. They effectively set the summation in (16) equal to its average times number of vehicles resulting in the following

$$q \cdot \frac{l_d+\bar{l}_v}{\bar{v}} = \sum_{i=1}^{q}\frac{l_d+l_i}{v_i} \qquad (17)$$

where, $\bar{l}_v$, is the mean effective vehicle length, and $\bar{v}$ is the average speed during the interval.

Substituting (17) into (16) the following relationship between spot speed, flow and occupancy may be obtained.

$$\bar{v} = q \cdot \frac{l_d+\bar{l}_v}{O \cdot \Delta t} \qquad (18)$$

It should be noted that in this setting average vehicle length is not directly measured. Normally, a constant average vehicle length is considered in the above equation. This is a biased estimator. Hall and Persaud (1989) proposed to adjust the estimator by multiplying a correction constant, however they pointed out that the effect of the bias is not uniform and a constant adjustment factor is not sufficient.

Dailey (1999) applied the Taylor's expansion up to the first two moments of the space-mean speed measurements, resulting in a non-linear function of the population speed parameter. This function was then linearized and used as the observation equation of a state-space model which was then solved by Kalman filter for population speed parameter. Ye et al. (2006) pointed out that the expansion approach is not robust and greatly depends on the linearization, the choice of initial guess and/or changes in vehicular speed. Ye et al. (2006) and Bickel et al. (2007) also use Kalman filter method to estimate vehicular speeds. Hazelton (2004) performed Bayesian analysis and applied Markov Chain Monte Carlo (MCMC) approach based on the assumption that speed in consecutive intervals follow a random walk. This method simulates the posterior distribution of vehicle speeds with a great improvement on accuracy; however, this offline approach is not practical for online estimation. Li (2009a,b) proposes a non-Gaussian Kalman filter and a recursive method for online vehicular speed estimation

Ahmed and Cook (1977) proposed a Box-Jenkins type model for flow and occupancy time series obtained from inductive loop detectors. Their model is essentially an ARIMA(0,1,3) model. They compared the performance of this model with three different smoothing algorithms; namely, moving average, double exponential smoothing, and exponential smoothing with adaptive response (Trigg-Leach method). However, they did not report on any modeling effort based on either travel time or speed data.

In a double loop setting, ILDs are able to provide an accurate estimate of vehicle speed based on the passage time lapse and distance between two loops. D'Angelo et al. (1999) proposed a non-linear time series to predict point speeds at the location of dual loop detectors on freeway segments in the short term. Then these point speeds are simply extended to an area from midpoint of the upstream segment to the midpoint of the downstream segment to evaluate travel times. Ishak and Al-Deek (2002) made a comprehensive analysis of this method. However, in their evaluation they did not use any ground-truth travel time or speeds. Instead, they compared predicted point speeds with observations from loop detectors. One of the major findings of Ishak and Al-Deek (2002) is that increasing rolling horizon (the duration of traffic evolution prior to current time used in predictions) would increase the relative error of travel time predictions. This is a counter-intuitive observation, since we expect a model should perform better when it uses more historical data as input. Additionally, they found that this method produces substantial errors under congested flow conditions. Relative errors of up to 30 percent are reported in less than 20mph range. In 20 to 50mph range errors are as high as 20 percent. Only, at free flow speeds higher than 50mph, relative errors are reported to be less than 5 percent.

Based on a simple shock wave analysis procedure and basic kinematic principle (8), Coifman (2002) proposed a method to build vehicle trajectories around the location of a dual loop detector placed in the middle or on either end of a basic freeway segment. These trajectories then can be used to estimate travel times on the freeway segment. Compared to the naïve travel time estimates such as (12) or (14), this method reduces the errors by almost half, but the average absolute error still remains at around 10 percent of the ground truth travel times. The accuracy of this method falls with increase in the length of the freeway segment under study. This method is based on the stationary assumption for traffic conditions all over the freeway segment and at all times. Therefore, under normal conditions where queues are formed and later dissipate, one detector depending on its location may not capture all the existing shock waves in the segment and time period of interest.

Sun et al., 2008 proposed a method based on interpolating point speeds read in three consecutive detector stations to estimate travel time on the segment between detectors. This method simply fits a quadratic speed trajectory on three point speeds at these detector stations. For any departure time at first station, it is not clear how one should determine the arrival times at two downstream stations; speeds for which are to be used in building the speed trajectory. The reported test case exhibits errors of up to 55% in travel time estimates using this method.

2.1.2.1 Non-linear filtering

Treiber and Helbing (2002) propose an adaptive smoothing method which is essentially a non-linear filter that transforms input stationary detector data into the smooth spatio-temporal functions. The non-linear filter is, in fact, an adaptive linear combination of two linear anisotropic low-pass filters each representing either free-flow or congested traffic status. Weight system in the linear filters is based on exponential functions with scaled relative space-time coordinates. The weight system in the upper combination level is a non-linear hyperbolic tangent function with bias toward congested traffic filter results. No quantitative measure for accuracy of travel time predictions using this method is given. However, visual evidence is given as to accuracy of estimations and predictions.

2.2 Models Based on Lagrangean Data

Lagrangean data is comprised of vehicle trajectory data obtained from tracing probe vehicles inside the road segment of interest. For this reason, this data type may also be called internal data. In this sense, full/partial vehicle trajectories and speed profiles are an example of such data. Full trajectory data is considered very rich since it basically provide a complete record of a vehicle movement and the speeds and travel time it has experienced. In general, trajectory data is both expensive and brings about a host of privacy issues. Therefore, in practice, internal data are still very rare even though GPS and cellphone tracking applications are becoming more popular as traffic data sources. To address some of the privacy issues, some cellphone companies are using virtual trip lines (VTL) concept to detect passage speed of a sample of vehicles at a set of pre-specified locations which amounts to Eulerian data similar to speed data collected at loop detectors. However, (near) real-time trajectory data as a bi-product of fleet management operations has become available in recent years in certain corridors. The latter provides a major source of Lagrangean traffic data at an affordable cost for travel time estimation for commercial purposes.

2.3 Models Based on Integrated Lagrangean Data

2.3.1 Automatic Vehicle Re-Identification (AVI)

Automatic license plate reader (ALPR), toll-tag readers and video processing systems capable of matching passing vehicles signatures between a pair of locations along the road are examples of these technologies. AVI data directly reflects realized travel times between two observation points, but at the same time it is more difficult to obtain compared to point measurements. Generally speaking, established traffic sensing technologies that are able to provide AVI data are both expensive and controversial in terms of exposing general public to privacy risks and therefore have found very limited geographical reach. As a result, earlier studies in this area tend to make use of widely available point sensors and to show that matching data from a pair of, for instance, loop detectors can result in accurate travel time estimates. In recent years, new emerging technologies have proved to be more effective in providing AVI data. Magnetic and Bluetooth matching sensors are examples of the new wave of AVI technologies.

Hoffman and Janko (1990) are the first who reported on implementing a travel time prediction system using AVI data. In their study, data was obtained from infra-red transmitter/receivers installed at over 230 signalized junctions and 10 locations on urban freeways in West Berlin. A small fleet of vehicles were equipped with the same infra-red capability as well as position finding devices so that their passage time in front of static devices could be recorded. Their proposed prediction methodology mainly consists of forming a historic data set and estimation of average travel time for each time interval, then a correction factor in the form of ratio of the observed travel time in the last interval to that same interval's historic average is used to predict current interval's travel time. Unfortunately, no measure of accuracy for this method is reported.

Dailey (1993) proposed a signature matching method for travel time estimation which uses cross-correlations between 5 sec vehicle counts from upstream and downstream inductive loop detectors at relatively short distances (0.5 mile is used in the reported example). In this method no effort is made to evaluate speeds from occupancy and therefore there is no need to estimate average length of vehicles. The method chooses the lag associated with the maximum cross-correlation value as the mean travel time between two consecutive detectors. The minimum acceptable cross-correlation value is reported as 0.4 which is shown to correspond to the 15 percent occupancy level. It is postulated that with increase in the congestion level beyond this point, the rigidity in flow of traffic between two points diminishes. Therefore, the method is not suitable for congested situations. No effort to validate the results of this method against ground-truth data has been reported.

Coifman and Ergueta (2003) propose an algorithm along with four separately designed filters to match signals between two consecutive dual loop detector stations on a single lane. The algorithm identifies a set of feasible upstream pulses for each downstream pulse; each pulse representing the passage of a vehicle. Then all vehicles in this set which have an estimated length range that includes that of the corresponding downstream vehicle are incorporated into a vehicle match matrix. Visual inspection of this matrix suggests that, under stable traffic conditions, correct matches should form a long vertical sequence of entries in the matrix. The method is therefore based on finding the longest vertical sequence. To eliminate false positives, four tests are introduced; filter test, cone test, travel time test and multiple lane change test. Results of a reported test study on two separate 0.55 km freeway segments demonstrate the accuracy of the method to be extremely good in comparison with ground truth. A mere 1.45 percent average absolute percent error in travel time is reported in a case where there is no on/off ramp between two detector stations. However, in a setting where an off ramp exists on the studied segment no error measure is reported.

Coifman (2003) considers the case of a pair of double loop detectors located at two ends of a freeway lane. In order to detect the start of congestion, he suggests that outstanding vehicle length estimates from downstream station be compared with length estimates from upstream station within a time window reflecting free flow travel times on the segment. If in consecutive time intervals such matches are not found then it is suggested that traffic is in congested mode. However, if a match is found then it provides a travel time estimate. This method works best when larger number of trucks (or any longer vehicles) is present in the mix. In his numeric test, Coifman managed to match 5% of traffic using this method. In its basic case, this may not be very valuable information since free-flow travel time is more or less a known constant (small variation). Therefore, he extends this method to the congested case by considering different travel speed ranges which results in a rudimentary method for travel time estimation under any traffic condition using existing point sensor technology. It should be noted that quantity of matches and also quality of estimates will decrease as congestion increases because during congestion more vehicles change lanes.

2.3.1.1 Time Series Analysis/(Non)Linear Filtering

Generally, methods falling in this category are based on signal processing ideas. It is conceived that travel time observations when ordered on the basis of the sequence of time intervals at which they have been measured provide a history of the evolution of a system.

$$Y = [y_t] \qquad (19)$$

Box, Jenkins, and Reinsel (1970) proposed statistical techniques to analyze time series. Auto

Regressive Integrated Moving Average (ARIMA) models provide a standard modeling framework in a typical time series analysis. The $ARIMA(p, d, q)$ model is expressed as,

$$(1 - \sum_{i=1}^{p-d} \varphi_i B^i)(1 - B)^d y_t = (1 + \sum_{i=1}^{q} \theta_i B^i)\varepsilon_t \qquad (20)$$

where,

p, is the order of auto-regressive terms,

d, is the number of sequential differencing needed to stationarize the time series

q, is the order of moving average terms,

φ, are the parameters of the auto-regressive part,

θ, are the parameters of the moving average part,

B, is the lag operator defined as $B^i(y_t) = y_{t-i}$, and

$[\varepsilon_t]$, are the error terms series assumed to be independent and identically distributed (i.i.d.) random normally distributed variables with mean equal to zero (white noise).

Dion and Rakha (2006) proposed a real-time adaptive exponential low-pass filtering algorithm for travel time estimation and prediction using very small sample AVI data (less than one percent of traffic volume) from toll-tag readers. They used toll-tag data from TransGuide system in San Antonio to demonstrate the method performance in predicting two minute time intervals. Aside from graphs, no other concrete measure of prediction accuracy is reported.

They assume that travel time is log-normal distributed. This algorithm uses a simple smoothing technique to forecast the future average and variance of travel time. The predicted average travel time is estimated according to the following equation,

$$\ln(\hat{y}_{t+k}) = \begin{cases} \alpha.\ln(y_t) + (1-\alpha).\ln(\hat{y}_t) & , k = 1 \\ \alpha.\ln(y_t) + (1-\alpha).\ln(\hat{y}_{t+1}) & , k = 2 \\ \alpha.\ln(\hat{y}_{t+k-2}) + (1-\alpha).\ln(\hat{y}_{t+k-1}) & , k > 2 \end{cases} \qquad (21)$$

where,

y_t, is the observed travel time at time interval t,

$\hat{y}_t$, is the estimated travel time at time interval t,

α, is the smoothing factor to linearly combine log-normal of travel times at time interval t, and

k, is the number of time steps ahead for which prediction is being performed.

Based on the predicted travel time average and variance a range for valid travel time observations in the next time interval can be specified. Observations that fall outside this validity window are dismissed as outliers. Essentially, in this method, specification of the validity range is performed based on the following four factors:

- Expected average trip time and trip time variability in future time interval,
- Number of consecutive intervals without any readings since the last recorded trip time,
- Number of consecutive data points either above or below the validity range, and
- Variability in travel times within an analysis interval.

2.3.1.2 State-Space Models

Chen and Chien (2001) use probe vehicle travel times as measurements in a trivial Kalman filter to predict travel times on a freeway path segment. They use historic travel time data to estimate transition parameter, $\phi(t)$, in the system model.

$$x(t) = \phi(t-1).x(t-1) + w(t-1) \tag{22}$$

$$z(t) = x(t) + v(t) \tag{23}$$

where,

$$\phi(t) = \frac{\hat{x}_H(t+1)}{\hat{x}_H(t)} \tag{24}$$

and, $\hat{x}_H(t)$ is the historic travel time associated with time interval t.

The CORSIM simulations are the source of their probe travel time measurements. They report maximum relative errors of 5 percent in their travel time predictions when probe vehicles represent 1% of traffic. Their prediction accuracy does not improve proportionally by increasing probe vehicles to 3% of traffic though.

Barcelo et al. (2009) propose a discrete Kalman filter (DKF) similar to Chen and Chien (2001) to estimate and predict travel times on a 40 km long freeway segment of AP-7 Motorway between Barcelona and the French border. However, the state transition function adopted in this work is set as the ratio of travel time estimates in the last two time intervals.

$$x(t) = \phi(t-1).x(t-1) + w(t-1) \tag{25}$$

$$z(t) = x(t) + v(t) \tag{26}$$

where,

$$\phi(t) = \frac{\hat{x}(t)}{\hat{x}(t-1)} \tag{27}$$

They used travel time measurements obtained from 6 Bluetooth vehicle re-identification sensors on each direction that were deployed anywhere from 4 to over 15 kilometers apart from each other. Raw travel time samples first have been filtered and aggregated in one minute time intervals. It is these one minute mean travel time estimates that are used in the DKF framework to predict travel times. Later, predictions are aggregated and reported in 5 minute time intervals. A very high correlation coefficient ($R^2 = 0.9863$) between the observed and predicted time series and a prediction MARE equal to 3.54% are reported. It should be noted that long distance and intercity nature of the data used to evaluate this method, to a large extent, would explain the high quality performance of this method in forecasting travel times. In this study, speeds below 70 km/h (45mph) are assumed to signal a congested condition which in itself reflects the high speed nature of operations on the facility under study.

2.4 Models Based on Eulerian and Integrated Lagrangean Data

In cases where both Eulerian data from two endpoints of the segment and travel time (integrated Lagrangean) observations between them are available then it is possible to investigate the relationship between the two data types. The effects of Eulerian data on travel time can be modeled and evaluated using Eulerian data as independent (descriptive) variables and travel time data as dependent variable. Essentially, in this setting, travel time can be modeled as an implicit/explicit function of the available Eulerian data.

$$y = f(\mathbf{x}) \tag{28}$$

When function $f(.)$ is not explicitly defined, inductive or statistical methods can be used to draw conclusions on the relationship between travel time and other Eulerian data. Non-parameteric models such as k-Nearest Neighbor (k-NN) are specifically of this type. On the other hand, when function $f(.)$ is assumed to take a linear form then linear regression models can be adopted to specify the relationship between travel time and the Eulerian data. However, in general, this relationship may be non-linear in nature. Therefore, general non-linear functions such as Artificial Neural Networks (ANN) may be used for this purpose.

Downside to these methods is that huge historic data sets are needed to calibrate the associated models. The results will highly depend on the extent of the historic data set and its representation of recurrent and non-recurrent traffic conditions. Moreover, these models tend to be site dependent, a property which limits the transferability of the estimated models.

2.4.1 Inductive/Statistical (Historic Data Based) Models

2.4.1.1 k-Nearest Neighbor Methods (k-NN)

These methods belong to the non-parametric category of travel time prediction methods. This implies that no assumption is necessary to be made on error distributions. Even though large historic data sets are necessary to make good predictions using k-NN, it is anticipated that over time historic data set gets richer and therefore performance of the k-NN method in predicting travel times is expected to improve. In this method, given the input vector $\mathbf{x}$, the following inference on the prediction error magnitude is made.

$$\|\mathbf{x} - \mathbf{x}_k\| \leq \varepsilon_k \implies \|y - y_k\| \leq \gamma_k \qquad k = 1, \ldots, K \tag{29}$$

where,

$\mathbf{x}_k$, is the k-th nearest neighbor to input vector ,

ε_k, is the measured distance between input vector $\mathbf{x}$ and its historic k-th nearest neighbor $\mathbf{x}_k$,

y_k, is the historic travel time associated with vector $\mathbf{x}_k$, and

γ_k, is the anticipated distance between predicted travel time y and its corresponding historic k-th nearest neighbor y_k.

Basically, equation (29) states that if input vector, $\mathbf{x}$, is close enough to its k-th nearest neighbor, $\mathbf{x}_k$, then its output, y, will be close enough to the historic output associated with the k-th nearest neighbor, y_k. Therefore, output y may be written as the sum of the k-th nearest neighbor's output, y_k, and (an unknown) function of the measured distance between input vectors, $g_k(\varepsilon_k)$.

$$y = y_k + g_k(\varepsilon_k) \qquad k = 1, \ldots, K \tag{30}$$

The output, y, then can be approximated as a function of all K nearest neighbor outputs.

$$y \cong h(y_1, y_2, \ldots, y_K) \tag{31}$$

Use of average function is a popular choice for function $h(.)$ in most circumstances.

$$y \cong \sum_{k=1}^{K} y_k / K \tag{32}$$

Handley et al. (1998) reported the first application of k-NN method to forecast travel times on a 25 mile southbound segment of I-5 in San Diego. The method takes into account weekday versus weekend, day of week, time of day, and the 30 second average traffic speeds reported from 116 loop detectors along this segment as four features based on which similarity between current conditions and historic observations are determined. In this application three nearest neighbors are selected and the average of their associated travel time is reported as predicted travel time for current time interval. This method resulted in a MARE of up to 20% during peak period and up to 7% during off-peak period.

Clark (2003) proposed a k-NN approach to forecast 10 minute time mean speeds from a set of loop detectors on the outer loop of London beltway M25. He used a set of four consecutive speed observations in the matching process to find 8 nearest neighbors in the historic database. The distance metric used in this study is the weighted sum of squares of distances between current and historic observations contributed from each parameter included in the analysis domain. Results show that speed forecasts will be best if only speeds are included in the process. A best MARE of 5% is reported for this method in predicting speeds 10 minutes ahead. However, based on the reported results, it seems that a naïve forecast (the current observation) will perform as well as the proposed method.

Robinson and Polak (2005) tested both isolated and combined effects the choice of distance metric, value of , and local estimation measure will have on the performance of k-NN method in forecasting an urban arterial travel times. Observed travel time data in this study are obtained using a pair of license plate matching cameras installed at two ends of a one kilometer long segment in central London. They found that k-NN method is not too sensitive to the choice of distance metric, and that a robust local estimation method is preferable to other methods. Also, they found that the optimal value of k depends on the size of the historical database. In their case study, a k-NN method with k equal to 2160 using standardized Euclidean with variance as weights for distance measurement and a locally weighted scatter plot smoothing (LOWESS) as estimation method was found to perform optimally. This method produced MARE equal to 18% in 15 minute travel time forecasts.

You and Kim (2000) reported on an early application of k-NN method on both freeway and arterial segments in Korea. The segments they studied, however, do not seem to reflect any serious congestion conditions. Similarly, Bajwa et al. (2005) report on an application of k-NN method on data from five long freeway segments in Tokyo metropolitan area. They reported RMSEP more than 10% for their applications in congested segments.

2.4.1.2 Linear Regression Methods

When function in equation (28) is assumed to be of linear type, then it can be written as follows.

$$y = A\mathbf{x} \tag{33}$$

where, A is the coefficients vector and can be estimated using linear regression methods such as least squares.

Kwon et al. (2000) propose a prediction method based on linear regression with stepwise variable selection. In their model historic travel time measurements are used as dependent variable against which flow and occupancy data from loop detectors are regressed as independent variables in a least squares error sense. They used data gathered on a 6.2 mile segment of I-880 south of Oakland, California for model evaluation. This data set includes measurements from double loop speed stations located at approximately one-third of a mile apart as well as probe travel time data (364 trips) from 20 weekday mornings. At 5 minute ahead, this linear regression method resulted in 9-15% MARE in travel time predictions. Obviously, this model is site specific and should be re-estimated for other segments using their corresponding data sets.

Zhang and Rice (2003) propose a time varying coefficient (TVC) linear model to improve upon a naïve predictor based on current speeds at two ends of a freeway segment. The method requires a large historic database to calibrate prediction model's coefficients. They reported on the method's performance on the north-bound direction of I-880 data set which was used by Kwon et al. (2000). While using historic dataset provides a slightly more than 10 percent MARE on travel time prediction, the TVC model has roughly 6% error in current travel time estimation and about 11% error at 30 minute forecasts.

Chakroborty and Kikuchi (2004) propose a simple linear regression model to estimate auto travel times based on measured bus running travel times. The latter is obtained using GPS devices installed on buses and is equal to total bus travel time minus times bus spends stopped at stations along the segment. The method evaluations on five arterial segments in northern New Castle County, Delaware revealed that in the worst case 77% of predictions made were within 10% of floating car measurements. In this study, there were 28 to 30 travel time measurements made at each site.

Liu and Chang, 2006 reported on attempts to calibrate single linear regression models to account for increase in travel time due to accumulations on segments with constant and variable capacity drops at the downstream. Data obtained from CORSIM micro-simulation runs is used to estimate the models. Applying the method in practice is difficult since model calibration requires a large historic database and the count data to be used in the method are not accurately available.

2.4.1.3 Artificial Neural Network (ANN) Methods

ANN is a general non-linear function approximation system that is inspired by generic functions of biological neural networks. The idea behind ANN is that data processing happens at many simple data processing units called neurons. Typically, in an ANN these neurons are organized in layers in a feed-forward network. Associated with each link in the network is a weight that should be determined using a training procedure such as error back-propagation. Input to each neuron is the weighted sum of outputs from neurons in the previous layer. Neurons act as a switch and depending on the input strength produce an output determined by an activation function. Identity, linear, binary step and sigmoid (S-shaped) functions such as logistic and hyperbolic tangent functions are among popular activation functions (Fausett, 1994). A feed forward ANN with L layers can be concisely represented as the following recursive equation.

$$y = f_L\big(\mathbf{w}'_{L-1,L} f_{L-1}\big(\ldots f_1\big(\mathbf{w}'_{0,1}\mathbf{x} + \mathbf{b}_1\big) \ldots + \mathbf{b}_{L-1}\big) + \mathbf{b}_L\big) \tag{34}$$

where,

$\mathbf{w}_{l,l+1}$, is the specified weight matrix between neurons in consecutive layers l, and $l+1$,

$\mathbf{b}_l$, is the bias vector in neurons of layer l, and

$f_l(.)$, is the vector of activation functions belonging to neurons of layer l.

Park and Rilett (1998) propose a clustering and artificial neural network method to forecast travel times on an urban freeway. They use AVI travel time data on link segments from just over one mile to 5 mile long on eastbound US-290. This is part of the automatic tolling system TranStar in Houston, Texas. Application of this method resulted in 5 minute travel time forecasts with over 8% MARE. Errors nearly doubled in 25 minute forecasts when MARE reached 16%.

Rilett and Park (2001) report on applying a spectral basis neural network to directly forecast freeway corridor travel times. In this method an extra layer is added to the front of ANN which implements Fourier transform. The transformed basis functions then will be used in a series of hidden layers to build a forecast for corridor travel times. Again, performance of the method on a 12.8 km segment of eastbound US-290 in Houston is reported. MAPE in 5 minute forecast has been about 6% while this same measure for 25 minute ahead forecasts has been more than 15% which is not that different from their older results.

In recent years, ANN methods with transformed input data have become more common place. Hamad et al. (2009) used Hilbert-Huang decomposition of the ILD speed signals as input to a speed predicting ANN. They tested this method on I-66 data. Their prediction MARE for 5 to 25 minute ahead during morning peak hour ranged from 6 to 10 percent.

2.4.2 Traffic Flow Theory Models

Traffic flow theory models can be categorized into two major groups. This classification is based on the level of detail at which a traffic stream is being modeled. Microscopic models track the movements of individual vehicles in traffic. These movements typically fall under two umbrella categories: car following and lane changing models. Microscopic models are computationally intensive and very difficult to calibrate and verify.

On the other hand, macroscopic traffic models deal with characteristics of a group of vehicles at an aggregate level. Variables such as flow and density are passage rate of vehicles at a cross section and their presence rate over a stretch of highway, respectively. Based on definition, these variables can be shown to be related through a third variable, namely space mean speed. This constitutive relationship along with an assumption on the form of dependence between speed and density leads to the so-called fundamental traffic diagram (FTD).

Macroscopic models are in fact conservation laws expressed in the form of partial differential equations (PDE). These models can be solved using exact methods such as method of characteristics. In real world applications, in general, it is difficult to obtain the exact solutions. Mesoscopic models approximate solutions to these conservation laws by breaking the solution domain into a series of smaller sub-domains. Finite difference (FD) methods such as up-winding and finite element (FE) methods such as Galerkin are typically used to approximate the evolution of traffic variables over time and space.

2.4.2.1 Microscopic Simulation Models

Liu et al. (2006) report on an online travel time prediction system customized for Ocean City, Maryland. The system receives data from 10 stationary sensors sparsely located along 30 miles of US-50 and MD-90 between Salisbury and Ocean City, Maryland. In their study they used a calibrated micro-simulation model based on CORSIM software package to predict travel times in the system. Forecast traffic volumes at detector locations needed for micro-simulation module are determined from a historic database using a nearest neighbor method. No specific measure of accuracy regarding predicted travel times is reported.

2.4.2.2 Mesoscopic Simulation Models

Waller et al (2007) adopt an ARIMA(3,1,2) to forecast inflows to the freeway segment under study, then they use a meso-simulation technique called cell transmission model (CTM) to simulate propagation and movements of vehicles inside the segment. Later, based on cumulative flow curves at the segment endpoints they are able to forecast travel time. On a 3 mile freeway segment, they reported 10 to 23 percent RMSE on travel times predicted 5 minutes ahead using this method when compared with travel times obtained from VISSIM micro-simulation.

2.4.2.3 Hybrid Models

Zou et al. (2007) propose a method for travel time estimation over long freeway segments. Their method is an extension of Coifman (2002), which makes use of occupancy and speed data from stationary detectors located at either end of the segment. They first identify different recurrent traffic patterns based on a historic data set. Then for each pattern, they calibrate a parameterized model to estimate travel times over the segment assuming that speeds at each detector can be extended using a linear relationship to represent the average travel speeds on each half segment. Later, based on a piecewise exponential speed-occupancy relationship and the assumption that traffic conditions at detectors will propagate with a constant speed within the segment, an iterative method for trajectory approximation is proposed. Performance of this method is reported in comparison with travel times obtained from vehicle re-identification conducted on two days' worth of video recordings at the endpoints of an over 10 mile long segment of I-70 between US-40 and I-695 east of Baltimore, Maryland. Results are counter-intuitive in the sense that the proposed method resulted in higher errors in free flow and heavy congestion conditions rather than in moderately congested periods. In free flow conditions errors up to 8.7% in travel time estimation are reported.

Yu et al. (2008) develop a hybrid model to predict travel times on a 7 mile long segment of US-50 leading to Ocean City, Maryland. They decompose travel time to a trend and a variation component. A fuzzy weighted average of clusters in the historic data base is used to estimate the trend term, while a cluster-based artificial neural network calibrated again on historic data base is used to predict travel time variations. Performance of the proposed hybrid model is compared with results obtained from micro-simulation software CORSIM. An average error of 8.7% in predicted travel times throughout the day is reported.

2.5 State Space Models

State space models provide a systematic general framework to represent the dynamics of the system no matter how complicated the system under investigation is. Additionally, it allows for incorporation of various measurements that become available dynamically over time into the estimation process. In general, a state space model consists of a system and a measurement equation. The idea is in some cases, it is difficult to make direct observations of a system state, instead it might be easier to observe and measure its correlated variables. Then the problem is to dynamically obtain best estimates of the system state by observing the correlated variables' evolution over time.

In its simplest form both the system and measurement equations in a state space model are linear. The following is an example of a discrete-time linear state space model:

$$v_{n+1} = M_n v_n + w_n \qquad (35)$$

$$y_n = H_n v_n + u_n \qquad (36)$$

where,

v_n, is the $N \times 1$ column vector of state variables at time step n

y_n, is the $M \times 1$ column vector of measured variables at time step n

M_n, is the $N \times N$ square transition matrix representing dynamics of the system at time step n,

H_n, is the $M \times N$ state to measurement transition matrix of the system at time step n,

w_n, is the $N \times 1$ column vector of state dynamics errors at time step n, and

v_n, is the $M \times 1$ column vector of measurement errors at time step n.

The best linear estimate of a state space model in the least square sense is obtained by Kalman filtering (Kalman, 1960 and 1961).

Nanthawichit et al., 2003 apply standard Kalman filter to the linearized approximation of a discretized version of the Payne's traffic model. The method primarily uses loop detector volume and speed measurements to estimate density and speed in a set of cells over time. Measurements from probe vehicles also may be included in this method in a very simplistic way. Average probe vehicle speed in each cell is regarded as measurement from an imaginary loop detector, if the cell in question does not include a loop detector. However, if a loop detector exists in the cell and we have two measurements from loop detector and probe vehicle in that cell at the same time interval, then the average of two measurements is used as measurement from the cell. Data generated by simulation software INTEGRATION has been used to evaluate methodology's accuracy. The suggested combined use of a traffic model along with stationary sensor and probe data in a Kalman Filter is shown to improve the travel time predictions by up to 36% compared to the autoregressive Kalman filter method proposed by Chen and Chien, 2001 which only uses probe data.

Sun et al., 2004 propose a Monte Carlo method based on a binary switching mode traffic model that only distinguishes between free-flow and congestion modes. In this method, first a fixed number of mode sample sequences with highest probability are identified, and then on each of these mode sample sequences a time varying Kalman filter is applied to estimate continuous traffic states (density). The a posteriori estimates of the continuous states are then computed as the weighted average of estimates from each Kalman filter. They use real stationary data from PeMS as well as simulation results from VISSIM to evaluate their method. They offer visual evidence that their proposed method is working well in estimating traffic mode; no quantitative measures are given though.

Chu et al, 2005 assume traffic flow and density are homogeneous on a freeway segment which may even include multiple on/off ramps. Also, they assume that all entering and exiting boundary flows to and from the segment are measured by means of stationary traffic sensors such as loop detectors, and they receive travel time measurements from probe vehicles traversing the segment every once in a while. An adaptive Kalman filter is proposed in which density is adopted as state variable and travel time measurements are simply related to the average density on the segment through a time-varying coefficient. A simple method for estimation of noise statistics (mean and variance of errors in system and measurement equations) based on an earlier work is given. Data generated using PARAMIC simulation at 30 second intervals on a 0.82 mile freeway segment with one on- and one off- ramp is used to evaluate the proposed method. They reported 8% mean relative errors in travel time estimates under recurrent morning peak conditions with a 5% probe rate, while under non-recurrent conditions (10 minute long incident blocking the right lane of freeway) this error measure is increased to about 10%.

Wang and Papageorgiou, 2005 report on using an extended Kalman filter to estimate density and speed on 500 meter freeway segments every 10 seconds. In their system equation a modified Payne-Witham model for dynamic speed estimation is used. Taylor series expansions are used to linearize the model equations at each point in time. Flow and speed measurements used in the estimation come from stationary traffic sensors at the boundaries of the freeway segment. Traffic data is generated using simulation based on the same traffic models as in the Kalman filter. In other words, Kalman filter is utilized o estimate traffic states given that we have full knowledge of actual traffic dynamics in the system. However, Kalman filter is used as a tool to identify the system state in presence of model and measurement noise. In their application, root mean square errors of the order 20% and 14% are reported for density and speed estimates, respectively. In the case of speed estimates, average absolute RMSE has been about 14 kilometers per hour (almost 9mph) on a 5 kilometer stretch of freeway. In a later work, Wang et al., 2007 used collected data from a 4.1 kilometer German highway to demonstrate the performance of their proposed methodology. In this case, no quantitative error measures for state estimates are reported.

Work et al., 2008 propose an ensemble Kalman filtering (EnKF) approach for highway traffic estimation in the presence of both stationary and probe vehicle data. They use a velocity based cell-transmission model (CTM-v) with a Greenshield's type fundamental diagram which makes it possible to work directly with measured speeds. They ran tests on a simulation model calibrated for I-880. This method resulted in 25% average relative error on speed estimates at 5% probe penetration rate.

Herrera and Bayen, 2009 use cell-transmission model (CTM) with a triangular fundamental diagram to estimate density given a combination of boundary and probe measurements. Newton relaxation and discrete Kalman filter are two methods that they used to estimate traffic conditions. They are using data collected in Next Generation SIMulation (NGSIM) project on US highway 101 in California, as well as GPS probe data from Mobile Century data collection effort on interstate 880 in California.

Claudel and Bayen, 2008 proposed a method based on viability theory in optimal control to estimate a lower and upper bound for the number of vehicles that are initially present on the road segment under investigation based on both stationary and probe data. Then, using the conservation of vehicles principle method is capable of estimating a range for travel time between the two end points of the segment. This method does not take into consideration presence of on and off ramps between the two end points. They tested their proposed method on US-101 dataset from NGSIM and I-880 from Mobile Century data collection effort. A mean relative error higher than 8% on travel time estimates is reported using this method.

Mihaylova and Boel, 2004 use a particle filter (PF) to estimate traffic variables from a nonlinear state-space model. PF is essentially a Bayesian recursive estimation approach analogous to a Monte Carlo simulation. Therefore, PF is computationally expensive but is most accurate in the case of nonlinear state-space models. Their numerical experiments on an undisclosed 0.5 kilometer long four lane freeway segment reported errors up to 10% in speed estimates.

Table 1 summarizes the distinctive features of relevant traffic state estimation studies reported in the literature which are particularly based on state-space models. Based on this table a few points should be noted. First, not many studies reported on the accuracy of travel time predictions. Second, as expected, time interval size plays a significant role in the accuracy of estimates. Third, travel time data has never been systematically incorporated into the estimation process. The only reported work that directly combines probe data as travel time with stationary data (Chu et al. 2005) does so through a simplifying assumption that travel time is an adaptive coefficient of density in the segment under study.

Table 1. Summary features of traffic speed/travel time estimation studies using state-space models.

Author(s)	Year	Traffic Model	Measurement(s)	Estimation Method	Data Source	Facility Type	Time Interval	Estimation Variable	Prediction Variable	Accuracy
Chen, Chien	2001		-Probe (travel time)	Auto-Regressive Kalman	CORSIM	Freeway (I-80 in NJ)	5 min	Travel Time	Travel Time (5 min)	MARE ~2% @ 1-3% probe
Treiber, Helbing	2002		-Stationary (flow, speed, density)	Adaptive Smoothing	Double ILD	Freeway (A8, A9, A5, Germany)	1 min	Density	Density (20 min)	Visual
Nanthawichit et al.	2003	Payne (linearized and discretized)	-Stationary (flow, speed) -Probe (speed) -Combined	Kalman Filter	INTEGRATION	Freeway (Yokohane, Japan)	10 sec est, 3 min pred.	Speed	-Speed -Travel Time	MARE<3-26% MARE <4% @ 3% probe
Mihaylova, Boel	2004		-Stationary (flow, speed, density)	Particle Filter	METANET		10 sec	Flow Speed Density		
Sun et al.	2004	SMM	-Stationary (flow, density)	Mixture Kalman Filter	PeMS (ILD) VISSIM	Freeway (I-210 in CA)	2 sec	Speed		Visual
Chu et al.	2005	LWR (discretized)	-Stationary (flow, density) -Probe (travel time) -Combined	Adaptive Kalman Filter	PARAMICS	Freeway (I-405 in CA)	30 sec			MARE ~10% @ 5% probe
Wang, Papageorgiou	2005	Modified PW (linearized)	-Stationary (flow, speed)	Extended Kalman Filter	Kalman Filter		10 sec	Density Speed		MARE <19-21% MARE ~14%
Wang, Papageorgiou	2007	Modified PW (linearized)	-Stationary (flow, speed)	Extended Kalman Filter	ILD	Freeway (A92, Germany)	10 sec	Flow Speed Density		Visual
Herrera, Bayen	2008	CTM (triangular flux)	-Stationary (density) -Probe (position, speed)	-Newton Relaxation -Kalman Filter	NGSIM Mobile Century	Freeway (US-101, I-880 in CA)	1.2 sec 8 sec	Density		?
Claudel, Bayen	2008	Moskowitz HJ PDE	-Stationary (density) -Probe (position, speed)	LP	NGSIM Mobile Century PeMS	Freeway (US-101, I-880 in CA)		Travel Time		MARE >8% @ 5% probe
Work et al.	2008	CTM-v	-Stationary (speed) -Probe (position, speed)	Ensemble Kalman Filter	PARAMICS Mobile Century	Freeway (I-880 in CA)	2 sec	Speed		MARE 25% @ 5% probe
Barcelo et al.	2009		-Probe (travel time)	Auto-Regressive Kalman	Pilot project	Freeway (AP-7 in Spain)	5 min	Travel Time	Travel Time	MARE 3.5%

2.6 Bayesian Filter for Traffic State Estimation and Prediction

Traffic state estimation and prediction are important for traffic surveillance and control. Travel time can be estimated or predicted afterward based on traffic state values (Chen et al. 2012, Chen, Rakha and Sadek 2011).

Since traffic states are usually not measured everywhere and measurement errors exist, traffic state estimation is necessary when dealing with local and noisy sensing data (Wang and Papageorgiou 2005, Wang, Papageorgiou and Messmer 2008). Alternatively, in the case of traffic state prediction, current traffic measurement data are used to forecast future traffic flow variables. Recently, the implementation of various traffic macroscopic models within recursive Bayesian filter approaches has been widely used for both traffic state estimation and prediction problems (Work. et al. 2008, Mihaylova, Boel and Hegyi 2007, Sau et al. 2007, Wang and Papageorgiou 2005, Wang et al. 2008, Cheng, Qiu and Ran 2006, Work et al. 2010). For each time interval, both the time update (prediction) and measurement update (estimation) processes are included in this framework. The sequence of the two processes within a single time interval categorizes the problem as data estimation or prediction. Once new measurement data are available, they are used to adjust the prior predicted value and obtain the estimation. Conversely, prediction is calculated by implementing the estimated value in the time update equation.

A framework with different combinations of macroscopic traffic models and Bayesian filtering technologies has been used to estimate or predict traffic state variables in recent years (Work. et al. 2008, Mihaylova et al. 2007, Sau et al. 2007, Wang and Papageorgiou 2005, Wang et al. 2008, Cheng et al. 2006, Work et al. 2010). There are two main advantages to this framework. Within the time update process, the relationship of traffic parameters across adjacent time intervals is accurately characterized by macroscopic traffic models. Apart from this, the recursive framework ensures that traffic state data are efficiently calculated using only data from previous states, not the entire history (Ristic 2004).

To construct this recursive Bayesian filter framework, a time series equation is needed to predict future traffic variables (i.e., flow [q], space mean speed [u], and density [k]) using current measured data. A macroscopic traffic system can be used to track the temporal and spatial dynamic traffic flow behavior along a freeway by constructing a time series traffic variable update equation (Wang and Papageorgiou 2005). Computing the traffic stream flow as the product of the traffic stream space mean speed and density reduces the problem to two independent variables that characterize the traffic stream state. If we assume that a fundamental diagram exists, then there is a unique relationship between traffic stream density and speed. Consequently, a time series equation used to predict a single traffic stream variable (the numerical solution of a first-order partial differential equation [PDE]) is required to estimate the three traffic stream variables. This is classified as a first-order macroscopic traffic model. Alternatively, if the fundamental diagram is not strictly enforced, a second equation is needed for the second traffic flow variable. The model is considered a second-order macroscopic traffic model by predicting two traffic variables (the numerical solution of two first-order PDEs).

Instead of a data-driven statistical approach, the time series equation derived from macroscopic traffic models has the advantage of describing the temporal and spatial dynamics of traffic flow behavior along a freeway based on physical principles. It has strong robustness and is easy to implement at various freeway locations. For instance, a modeling algorithm proposed in (Mihaylova et al. 2007) uses sending and receiving functions to represent the traffic perturbation behavior of propagating forward and backward. Both traffic flow and speed are used in the state variable to construct a second-order macroscopic model. A similar second-order macroscopic model was proposed in (Cheng et al. 2006) using the concept of handoff, which is a mechanism that transfers an ongoing call from one cell to another when a cell phone user moves through the coverage area of a cellular system. Sending/receiving functions represent the vehicles that intend to leave/enter a segment. A more popular approach with which to derive time series equations is the traditional Lighthill-Whitham-Richards (LWR) model, which ensures that vehicle conservation is maintained on the road. For instance, the conservation equation in (Sau et al. 2007) is directly derived from LWR, and the change of section traffic flow for each time interval is constrained by the freeway section flow supply and demand. Both freeway density and flow volume are the state variables and are measured using loop detectors to estimate traffic state and then predict travel time. The conservation equation is also included in a second-order macroscopic traffic flow model and is used in the prediction process computation for traffic state estimation in (Wang and Papageorgiou 2005, Wang et al. 2008). Loop detector data are also used for measuring the state variables of traffic flow and speed. Within the application of using speed data from mobile devices to estimate the freeway traffic state, a velocity cell transmission model (CTM-v) is derived from the LWR by replacing the traffic flow and density with traffic speed based on Greenshields' fundamental diagram (Work. et al. 2008). Follow-up research demonstrates that the solution of the new PDE is equivalent to the LWR PDE under a quadratic flux function (Work et al. 2010).

After obtaining the time series equation, a recursive Bayesian approach is used to incorporate the measurement data to update state variables from the time series equation. The classical KF is the easiest way to incorporate the error between state prediction and measurement data for estimation purposes. However, the classical KF works ideally only for linear systems with Gaussian noise. Since most of the derived time series equations are characterized by nonlinear behavior, an extended Kalman filter (EKF) is widely used for traffic state estimation (Wang and Papageorgiou 2005, Wang et al. 2008). EKF is a revised classical KF with the calculation of Jacobian expression. This method has the same advantage as the classical method of propagating the error covariance matrix but can deal with a nonlinear system using Tyler estimation. However, it is difficult to compute the Jacobian expression for many nonlinear time series equations. By overcoming the defect of Jacobian computation and producing ideal accuracy for a nonlinear estimation, Ensemble Kalman filter (EnKF) enables the use of nonlinear evolution equation while exploiting the linear observation equation. EnKF uses Monte Carlo integrations to maintain the nonlinearity of error statistics. It has the same feature as KF that propagates errors by Kalman gain (Work. et al. 2008) and provides the average estimation or prediction output. However, it cannot deal with the model of nonlinear measurement equation and cannot output reliability state information. Regarding these problems, a powerful approach named a particle filter is implemented, which is applicable for any nonlinear system of equations and has no requirement for the distribution of the system noise (Mihaylova et al. 2007). A particle filter provides another benefit in that it delivers the estimation and prediction results as a distribution instead of just one value (Ristic 2004).

Although many studies have been conducted using different combinations of macroscopic traffic models and filtering techniques, these approaches suffer from a number of deficiencies. First, simple traffic stream models may not accurately replicate field-observed traffic stream characteristics. For instance, the Greenshields model does not provide sufficient degrees of freedom to replicate field observations (Rakha and Crowther 2002). As a result, the traffic state time series equation derived from simple traffic stream models usually results in lower estimation or prediction accuracy. On the other hand, a complex model (e.g., second-order macroscopic model) has many parameters; thus, the calibration of the model becomes a challenge (Cheng et al. 2006, Mihaylova et al. 2007, Wang and Papageorgiou 2005, Wang et al. 2008). Third, although deriving the time series equation from the LWR based on vehicle conservation law is a promising approach, no underlying traffic stream models other than the Greenshields model have been tested or evaluated. Furthermore, although these methods can be used for traffic state prediction in terms of the features of recursive Bayesian filters, few prediction results are presented or compared. Since predicting future traffic state data has many realistic applications for ramp metering, incident detection, and travel information broadcasting (Sau et al. 2007), it is necessary to conduct research about traffic state prediction.

A particle filter approach was proposed in order to solve the mentioned problems of predicting traffic state (Chen et al. 2011, Chen et al. 2012). This approach combines a more realistic traffic stream model – Van Aerde model with the traditional LWR flow continuity equation to derive a new time series equation to describe the spatial and temporal relationship of traffic speed on a freeway stretch. After implementing the speed equation in the particle filter framework, a multi-step prediction approach was proposed and tested, and was demonstrated to produce accurate traffic state predictions with less than 5% errors for a 5-minute prediction horizon (Chen et al. 2011). The effect of ramp flow and a more realistic boundary condition are considered in the following research and the testing results indicate that the proposed approach produces half of the prediction errors as compared to the LWR formulation in terms of traffic speed and density (Chen et al. 2012).

The above approaches utilize the macroscopic traffic model and the Bayesian filter to increase the accuracy of predicting traffic state in the near future. However, the accuracy degrades rapidly with the increase in the prediction time span (Chen et al. 2012, Chen et al. 2011). It should be noted that traffic state in the near future usually cannot provide enough information to cover the entire trip, especially for long trips. For instance, in the case of a 100-mile trip, departures at the current time would still be traveling 1 hour in the future even under free-flow traffic conditions. For this case, the traffic state for the following 1 hour or more should be predicted in order to compute dynamic travel times. An alternative to solving this problem is to use historical data. The historical data set provides a pool of past experienced traffic patterns which can be used to predict future traffic states. The key issue is determining the similar historical traffic patterns to match up with the changeable real-time traffic information.

2.7 Dynamic Travel Time Prediction

Travel time prediction is an essential part of an Advanced Traffic Management System (ATMS) and Advanced Traveler Information System (ATIS). The Federal Highway Administration (FHWA) encourages all Traffic Management Centers (TMCs) to post travel time and incident information, which not only provides useful information to motorists but also assists them in planning their route choices. This planning can cause a small number of drivers to divert away from the congested highway, thus providing critical additional capacity and assisting in the management of congestion (Chu 2011).

Various traffic sensing technologies have been used to collect traffic data for use in computing travel times, including point-to-point travel time collection (e.g., license plate recognition systems, automatic vehicle identification systems, mobile, Bluetooth, probe vehicle, etc.) and station-based traffic state measuring devices (loop detector, video camera, remote traffic microwave sensor, etc.). Private companies such as INRIX integrate different sources of measured data to provide section-based traffic state data (speed, average travel time), which is used in our study to develop algorithms for predicting travel times. The benefit of using section-based traffic state data is that travel time can be easily calculated from traffic state data. More importantly, the section-based data provide the flexibility for scalable applications on traffic networks.

By providing section-based traffic state data, there are two approaches to compute travel time depending on the trip experience (Tu 2008, Mazare et al. 2012). Dynamic travel time is the actual, realized travel time that a vehicle could experience during a trip. If a vehicle leaves its origin at the current time, the roadway speed will not only change across space but also across time during the entire trip. Consequently, dynamic travel time can be obtained by using a prediction algorithm to compute the speed evolution in future time steps. Instantaneous travel time is the other approach available to compute travel times without the consideration of speed evolution across time. It is usually computed using the current speed along the entire roadway; in other words, the speed field is assumed to remain constant in time. The instantaneous travel time is close to the dynamic travel time when the roadway speed does not change significantly across time axles during the trip. However, this approach may deviate substantially from the actual, experienced travel time under transient states during which congestion is forming or dissipating during a trip (Chen and Rakha 2012).

During past decades, many studies have been conducted to predict travel times. Some of the reviews of different methods can be found in earlier publications (Du, Peeta and Kim 2012, Myung et al. 2011, Lint, Hoogendoorn and Zuylen 2005, Vlahogianni, Golias and Karlaftis 2004). According to the manner of modeling, those methods can be classified into time series models, including: Kalman filter (Fei, Lu and Liu 2011, Yang 2005), Auto-Regressive Integrated Moving Average (ARIMA) models (Xia, Chen and Huang 2011, Xia and Chen 2009, Yang 2005) and data-driven methods, such as neural networks (Hinsbergen et al. 2011, Lint et al. 2005), support vector regression (SVR) (Vanajakshi and Rilett 2007, Wu, Ho and Lee 2004) and K-Nearest Neighbor (KNN) (Qiao, Haghani and Hamedi 2012, Myung et al. 2011, Bustillos and Chiu 2011) models. These techniques are implemented through direct and indirect procedures to predict travel times using different types of state variables. Travel time is directly used as the state variable in model-based or data-driven methods to predict travel times. Indirect procedures are performed by using other variables (such as traffic speed, density, flow, occupancy, etc.) as the state variable to predict traffic status, and then future travel time can be calculated based on the transition to predicted traffic status.

Time series models construct the time series relationship of travel time or traffic state, and then current and/or past traffic data are used in the constructed models to predict travel times in the next time interval (Yang, Liu and You 2010). Kalman filter is a popular method for data estimation and tracking, in which time update and measurement update processes are included. A time series equation is used to predict state variables and then state values are corrected according to the new measurement data. The main advantage of KF is that the recursive framework ensures that traffic data is efficiently updated using only data from previous states and not the entire history (Chen et al. 2012). Kalman filter methods were proposed to predict travel times using Global Positioning System (GPS) information and probe vehicle data (Yang 2005, Nanthawichit, Nakatsuji and Suzuki 2003). The state transient parameter in the time series equation is defined from average historical data to calculate future travel times. A similar idea was used in the Bayesian dynamic linear model for real-time short-term travel time prediction (Fei et al. 2011). The system noise can be adjusted for unforeseen events (incidents, accidents, or bad weather) and integrated into the recursive Bayesian filter framework to quantify random variations on travel times. The experiment results based on loop detector data from a segment of I-66 demonstrates that the proposed method produces higher prediction accuracy under both recurrent and non-recurrent traffic conditions. However, in these methods a problem exists in that the travel time in the previous time interval is needed to calculate the future travel time. For real-time applications, the travel time is usually greater than the time interval step size. Hence, the actual travel time from the previous time interval is not available to apply in the algorithms used to predict travel times for the next time interval.

A seasonal ARIMA model was proposed to quantify the seasonal recurrent pattern of traffic conditions (occupancy) (Xia et al. 2011, Xia and Chen 2009). Moreover, an embedded adaptive Kalman filter was developed in order to update the occupancy estimate in real-time using new traffic volume measurements. Consequently, multi-step look-ahead occupancy information is estimated to obtain a data matrix representing the temporal-spatial traffic conditions for the future trip. Since travel time cannot be directly computed through traffic conditions (occupancy), future traffic speed can be calculated using occupancy data by assuming an average vehicle length and using a constant conversion factor known as the g-factor in the literature. Consequently, dynamic freeway corridor travel times are predicted with the consideration of traffic state evolution along the corridor. However, this approach may be difficult to implement since the described recurrent pattern of traffic conditions may not be found everywhere.

Data-driven methods usually predict travel times using a large amount of historical traffic data. Time series models are not specified in the data-driven methods, considering the complex randomness of traffic systems. Neural networks can be trained from historical data to discover hidden dependencies which can be used for predicting future states. A state-space neural network (SSNN) method was proposed to predict freeway travel times for missing data (Lint et al. 2005). The missing data problem was tackled by simple imputation schemes, such as exponential forecasts and spatial interpolation. Travel time was the direct state variable used for the training process and the experiment results demonstrated that the SSNN methods produced accurate travel time predictions on inductive loop detector data. Supported vector machine (SVM) is a successor to artificial neural network (ANN); it has greater generalization ability and is superior to the empirical risk minimization principle as adopted in ANN (Wu et al. 2004). The application of SVM to time series forecasting is called support vector regression (SVR). The SVR predictor was demonstrated to perform well for travel time prediction. The point-to-point travel time is usually used as the input to ANNs and SVRs. However, both methods require long training processes and are nontransferable to other sites (Myung et al. 2011).

The KNN method can be used to find several candidate sequences from historical data by matching current with short past data sequences. Travel time and occupancy sequences were used to predict dynamic travel times using the KNN method with combined data from vehicle detectors and automatic toll collection systems (Myung et al. 2011). The occupancy was used since travel time sequence was collected for the recent past time intervals. The results from the case study demonstrated the improvement of prediction accuracy by combining two types of sequences for the matching process. Moreover, a KNN method was proposed by selecting candidates through the Euclidean distance and data trend measures to predict freeway travel times under different weather conditions (Qiao et al. 2012). Unlike ANNs and SVRs, KNN methods are easy to implement at different sites without data training required.

In summary, existing methods are either insufficient or have limitations in predicting dynamic travel times for departures at the current time or sometime into the future. The proposed approach used in this study is a data-driven method, yet it outperforms the previous methods by fully utilizing the relationship between traffic states and travel times. Moreover, unlike previous studies using travel time sequences as input, the proposed method uses temporal-spatial traffic data to match traffic patterns between real-time and historical data. Many advanced pattern matching techniques can be implemented in the proposed approach to find similar historical traffic patterns more efficiently and accurately, and obtain better travel time prediction results.

3 Methodology of Traffic Estimation

In this chapter solution methods for the proposed problem and its variants are presented. The proposed approach is based on the notion that in order to make quality predictions for the future of a system first we must have a good estimate of current state of the system and then utilize means to predict the system evolution in the future. Therefore, the work horse of such estimation and prediction framework is an accurate traffic model that is capable of reproducing any significant changes in traffic stream. The next issue that has to be addressed is that the best use of available data has to be made. Incorporating some traffic data types such as speed into estimation and prediction framework is straight-forward, while no systematic approach to fuse travel time data is currently available. Last, but not least, piece of the proposed approach has to do with drawing the best estimates out of the model and available data in presence of modeling and measurement errors and other real life constraints such as missing, intermittent and or out of sequence measurements.

In this chapter, first a well-known first order continuum traffic flow model is adopted to represent the dynamics of the system. An equivalent form of this model in terms of speed is derived. This model provides a theoretical framework to understand and analyze traffic processes on a variety of roadway facilities. Also, this model is widely used in an array of traffic operation and control applications worldwide. Therefore, it is essential to have both efficient and accurate solution methods for this model. A finite difference and a finite element method for solving the velocity based equivalent of the first-order continuum traffic flow model numerically are proposed.

Second, some desirable properties of travel time which are adopted in this study are briefly presented. Enforcing these properties on the estimates defines a feasible solution region for travel time partial derivatives. Then, interpreting travel time as distance to a boundary in space-time domain we introduce a framework to relate integrated lagrangean (travel time) data and local speeds. Relevant derivations are presented. Finite difference schemes to solve for travel times given speeds are introduced.

Finally, to derive the optimal estimates from the resulting state space model in presence of errors in modeling and measurements we propose optimal filtering approach. Kalman filtering (H_2), H_∞ and their extensions for nonlinear models and measurement equations are introduced. In particular, extended Kalman filtering (EKF), unscented Kalman filtering (UKF) and particle filtering (PF) and their H_∞ equivalents are discussed. Methods to address missing and out of sequence measurements are introduced.

3.1 Traffic Model (LWR-v)

Continuum traffic flow theory is a powerful tool to describe the evolution of macroscopic traffic parameters over time and space. This is in contrast to microscopic models of traffic flow which generally require meticulous handling of individual vehicles movements in the traffic stream. The most basic continuum traffic flow theory builds on two basic physical principles that is conservation of vehicles and the fundamental relationship between flow rate, density and speed. Conservation principle states that no vehicle is added or lost in traffic at any time other than the ones that enter or exit through the boundaries. This basic continuum theory was first proposed by Lighthill and Whitham (1955) and Richards (1956). Despite its simplicity, and therefore its inherent limitations (Daganzo, 1997), the so called kinematic wave theory or LWR model provides a good approximation to the dynamics of traffic flow which has proved to be useful for most practical purposes.

Even though our ability to directly measure different traffic parameters has dramatically increased over the years, the measurements are still widely different in terms of their accuracy and reliability. For instance, flow rate and density both can be obtained as a result of simple counting processes performed at a single point or a pair of points, respectively. However, it is ironic that in practice, large inconsistencies between counts in consecutive stations (with no exit or entrance in between) exist. In the case of loop detectors this "drift" phenomenon is well known. In addition, at a macroscopic level, definition of density is a bit ambiguous in the sense that the length over which concentration of vehicles would affect a driver's behavior under normal conditions is not specified. In contrast, speed measurements which theoretically are expected to be more difficult to obtain have proved to be a far more reliable source of traffic data. That is why, in this study, we focus on a speed-based equivalent of the LWR model.

The rest of this section is organized as follows. First, we briefly present the derivation of the first-order speed based continuum traffic flow model based on LWR model. Second, for the sake of completion, we present the speed based finite difference method equivalent to cell-transmission model presented by Work et al. (2008). Third, a finite element solution method for the speed based LWR model is derived.

3.1.1 LWR-v DERIVATION

The first-order continuum traffic flow model proposed by Lighthill, Whitham and Richards (LWR) is considered in its differential from:

$$\frac{\partial k(x,t)}{\partial t} + \frac{\partial q(x,t)}{\partial x} = 0 \tag{1}$$

where,

k(x, t) is the traffic density at time t and at point x along the highway, and

q(x, t) is the traffic flow rate at time t and at point x along the highway.

This is the conservation law which essentially implies that no vehicle is born or lost along the highway. This equation is originally borrowed from hydrodynamics but has found widespread use in modeling vehicular traffic flow despite obvious differences between the two fields. In order to solve this equation, it is assumed that flow is a function of density and therefore we can write:

$$q(x,t) = Q(k(x,t)) \tag{2}$$

where, Q(.) is the flux function or what is called the fundamental diagram in traffic flow theory. One of the most important flux functions is the one proposed by Greenshields (1935) which suggests that a linear relationship exists between speed, v and density k:

$$\frac{v}{v_f} + \frac{k}{k_j} = 1 \tag{3}$$

where, v_f, is the free flow speed, and k_j is the jam density of highway under prevailing conditions.

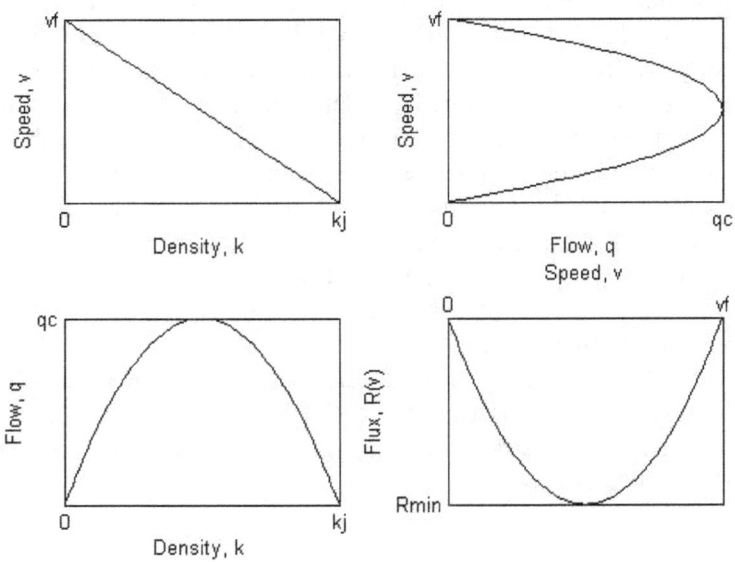

Figure 3. Greenshield's speed-flow-density relationships.

Figure 3 depicts the idealized relationship between pairs of traffic parameters under stationary conditions as hypothesized by Greenshields (1935). Also, under stationary traffic conditions, by definition the following relationship between flow, speed and density holds:

$$q = k\, v \tag{4}$$

Thus plugging the value of density from (3) into (4), we obtain the following quadratic relation between flow and speed:

$$q = k_j \left(v - \frac{v^2}{v_f}\right) \tag{5}$$

Likewise, plugging the value of density from (3) and flow from (5) into equation (1), we obtain the speed-based LWR model (LWR-v):

$$\frac{\partial v}{\partial t} + \frac{\partial R(v)}{\partial x} = 0 \tag{6}$$

where the new flux function is given by

$$R(v) = v^2 - v_f\, v \tag{7}$$

Bardos et al. (1979) have shown that partial differential equations of type (6) with the initial condition (8) and the weak boundary conditions (9) and (10) in space $]a, b[\, \times\,]0, T[$ are well-posed.

$$v(x, 0) = v_0(x) \quad , \forall x \in\,]a, b[\tag{8}$$

$$\begin{cases} v(a, t) = v_a(t) \text{ or} \\ R'(v(a, t)) \leq 0 \text{ and } R'(v_a(t)) \leq 0 \text{ or} \\ R'(v(a, t)) \leq 0 \text{ and } R'(v_a(t)) \geq 0 \text{ and } R(v(a, t)) \geq R((v_a(t)) \end{cases} \tag{9}$$

$$\begin{cases} v(b, t) = v_b(t) \text{ or} \\ R'(v(b, t)) \geq 0 \text{ and } R'(v_b(t)) \geq 0 \text{ or} \\ R'(v(b, t)) \geq 0 \text{ and } R'(v_b(t)) \leq 0 \text{ and } R(v(b, t)) \geq R((v_b(t)) \end{cases} \tag{10}$$

In the following two sections, a finite difference and a finite element method are proposed to numerically solve LWR-v model (6-10).

3.1.2 FINITE DIFFERENCE METHOD (FDM)

In practice, the LWR-v model derived in the previous section has to be approximated. In this section a finite difference approximation to LWR-v model proposed by Work et al. (2008) is briefly presented. This solution is in fact a velocity cell transmission model (CTM-v) similar to the ordinary CTM proposed by Daganzo (1994) which provides an approximate solution to LWR model. CTM-v follows a Godunov numerical scheme in which both time and space dimensions are discretized.

First, time is divided into N time intervals $\{t_n | n = 0,1, ..., N\}$ each of length $\Delta t = T/N$ and space is divided into M space cells $\{x_i | i = 0,1, ..., M\}$ each of length $\Delta x = (b - a)/M$. To each space cell x_i at time interval t_n, a discrete average speed v_n^i is assigned. Speed evolution at each space cell over time is given by,

$$v_{n+1}^i = v_n^i - \frac{\Delta t}{\Delta x}\left(g(v_n^i, v_n^{i+1}) - g(v_n^{i-1}, v_n^i)\right) \tag{11}$$

where the flow g is defined as,

$$g(v_1, v_2) = \begin{cases} R(v_2) & \text{if } v_1 \leq v_2 \leq v_c \\ R(v_c) & \text{if } v_1 \leq v_c \leq v_2 \\ R(v_1) & \text{if } v_c \leq v_1 \leq v_2 \\ \max(R(v_1), R(v_2)) & \text{if } v_1 \geq v_2 \end{cases} \tag{12}$$

where v_c is the minimum of the convex flux function (7). So, in the case of Greenshield's flux (7), we will have $v_c = \frac{v_{max}}{2}$.

To ensure stability of the numerical discretization method, the length of both spatial and time steps must meet the Courant-Friedrichs-Lewy (CFL) condition:

$$\left|\frac{\Delta t}{\Delta x} \max(R'(v))\right| \leq 1 \tag{13}$$

The CFL condition suggests that no entering vehicle at the beginning of a time interval can exit the cell in that same time interval. Thus, in the case of Greenshield's flux function, this condition translates to

$$\Delta t \times v_{max} \leq \Delta x \tag{14}$$

3.1.3 FINITE ELEMENT METHOD (FEM)

In this section a finite element solution method to the LWR-v PDE on a homogeneous highway is derived. Elements used for this purpose are one-dimensional simplex elements and approximation inside the element is performed using a linear equation. For a given element (e) extending from x_i to x_j, where $j = i + 1$ and with the length $l^{(e)} = x_j - x_i$, the interpolation function can be written as

$$v^{(e)}(x) = \begin{bmatrix} 1 - \frac{x-x_i}{l^{(e)}} & \frac{x-x_i}{l^{(e)}} \end{bmatrix} \begin{bmatrix} v_i \\ v_j \end{bmatrix} = [N(x)] \cdot \vec{v}^{(e)} \tag{15}$$

where, $[N(x)]$ denotes the vector of shape functions, and $\vec{v}^{(e)}$ is the vector of unknown nodal speeds in an element. Applying the Galerkin's method to the PDE leads to solving the following set of non-linear simultaneous equations. It should be noted that $\alpha = v_j - v_i$ is the difference of unknown speeds at the two end nodes of a given one-dimensional element.

$$\begin{bmatrix} 2(l^{(e)} + 2\alpha\Delta t) + 3v_{max}\Delta t & (l^{(e)} + 2\alpha\Delta t) - 3v_{max}\Delta t \\ (l^{(e)} + 2\alpha\Delta t) + 3v_{max}\Delta t & 2(l^{(e)} + 2\alpha\Delta t) - 3v_{max}\Delta t \end{bmatrix} \cdot \vec{v}^{(e)}_{t+\Delta t} = \begin{bmatrix} l^{(e)}\left(2v_i(t) + v_j(t)\right) \\ l^{(e)}\left(v_i(t) + 2v_j(t)\right) \end{bmatrix} \quad (16)$$

This represents the element equations where the left hand side matrix is the element characteristic matrix $\left([S^{(e)}]\right)$ and the right hand side vector is the element characteristic vector $\left(\vec{P}^{(e)}\right)$. This would allow us to rewrite (16) in the following standard form

$$[S^{(e)}] \cdot \vec{v}^{(e)} = \vec{P}^{(e)} \quad (17)$$

Therefore, the system equations can be assembled using element equations for each node in the highway system

$$[S] \cdot \vec{v} = \vec{P} \quad (18)$$

where,

$$[S] = \sum_{e=1}^{E} [S^{(e)}] \quad (19)$$

$$\vec{P} = \sum_{e=1}^{E} \vec{P}^{(e)} \quad (20)$$

However, it should be noted that these summations do not indicate the usual algebraic operation they are commonly used for, but they indicate assembly over finite elements. Essentially, in the assemblage process of system matrix and system vector from element matrices and vectors, care should be exercised to satisfy all the compatibility requirements at the element nodes. Boundary conditions also may be incorporated into set of system equations (18) before solving them (Rao, 2002). Newton-Raphson method may be used to solve the non-linear set of simultaneous equations (18). The following algorithm may be used iteratively at every time interval n to solve for the vector of nodal speeds, $\vec{v}_{n+1}$, in the next time interval.

Step 0 (initialization): Start with an initial solution for the next time step, $\vec{v}^{(0)}_{n+1}$, and evaluate the initial system function vector, $\vec{F}^{(0)}_{n+1} = [S^{(0)}_{n+1}] \cdot \vec{v}^{(0)}_{n+1} - \vec{P}_{n+1}$. Set the iteration counter j = 0. Set the convergence criteria, ε.

Step 1 (stopping criterion): If $\left|\vec{F}^{(j)}_{n+1}\right| \leq \varepsilon$ then stop. Accept $\vec{v}^{(j)}_{n+1}$ as the solution. Otherwise, continue.

Step 2 (function Jacobian evaluation): Evaluate the Jacobian, $[J^{(j)}_{n+1}]$, of $\vec{F}^{(j)}_{n+1}$ with respect to $\vec{v}^{(j)}_{n+1}$.

$$[J^{(j)}_{n+1}] = \left[\frac{\partial \vec{F}^{(j)}_{n+1}}{\partial \vec{v}^{(j)}_{n+1}}\right] \quad (21)$$

Step 3 (linearized approximation): Solve the following linearized approximation to the system function vector to find the vector of speed changes, $\overrightarrow{\Delta v}_{n+1}^{(j)}$.

$$\left[J_{n+1}^{(j)}\right] \cdot \overrightarrow{\Delta v}_{n+1}^{(j)} = -\vec{F}_{n+1}^{(j)} \tag{22}$$

Set modified vector of nodal speeds as $\vec{v}_{n+1}^{(j+1)} = \vec{v}_{n+1}^{(j)} + \overrightarrow{\Delta v}_{n+1}^{(j)}$. Set $j = j + 1$.

Step 4 (function evaluation): Evaluate the system function vector, $\vec{F}_{n+1}^{(j)} = \left[S_{n+1}^{(j)}\right] \cdot \vec{v}_{n+1}^{(j)} - \vec{P}_{n+1}$. Go to step 1.

3.2 Integrated Lagrangean (Travel Time) Data Representation

Given the fact that travel time of a vehicle is the line integral of inverses of its speeds along its trajectory, it is clear that either we should have the trajectory or the inverse problem of finding speeds based on travel time is in fact under-determined.

$$\tau = \int \frac{1}{\sqrt{1+v(X(s),s)^2}} ds \tag{23}$$

On the other hand, given a speed field, constructing the trajectory and evaluating travel time is a direct problem which is, at least theoretically, a straight-forward process. However, in practice this process is very inefficient since we should numerically approximate the above line integral. Also, travel time estimates obtained using trajectory construction methods will have a poor quality since errors in speed estimates will be accumulated in this process. Instances of this method resulted in errors up to 10 percent in travel time estimates over a half a mile segment. More details on these numerical experiments are given in the next chapter.

Here instead of the integral representation, let's focus on local variations in travel time. In other words, differential equations relating travel time and speed seem to be more useful in this setting. In this context we should redefine the travel time as the minimum distance from any given point (x, t) in the solution domain to the downstream boundary $(L, .)$. This definition along with the assumption of smoothness of travel time will result in a simple representation of travel times in the space-time solution domain as illustrated in Figure 4. In Figure 4 points in the space-time domain with the same travel time to the downstream boundary are shown in the form of a set of iso-distance contours. Also, this representation suggests that travel times at upstream (or any other point along the highway) is in fact a cross-section of various contours. Partial differential equation schemes may be used to numerically solve the proposed travel time model.

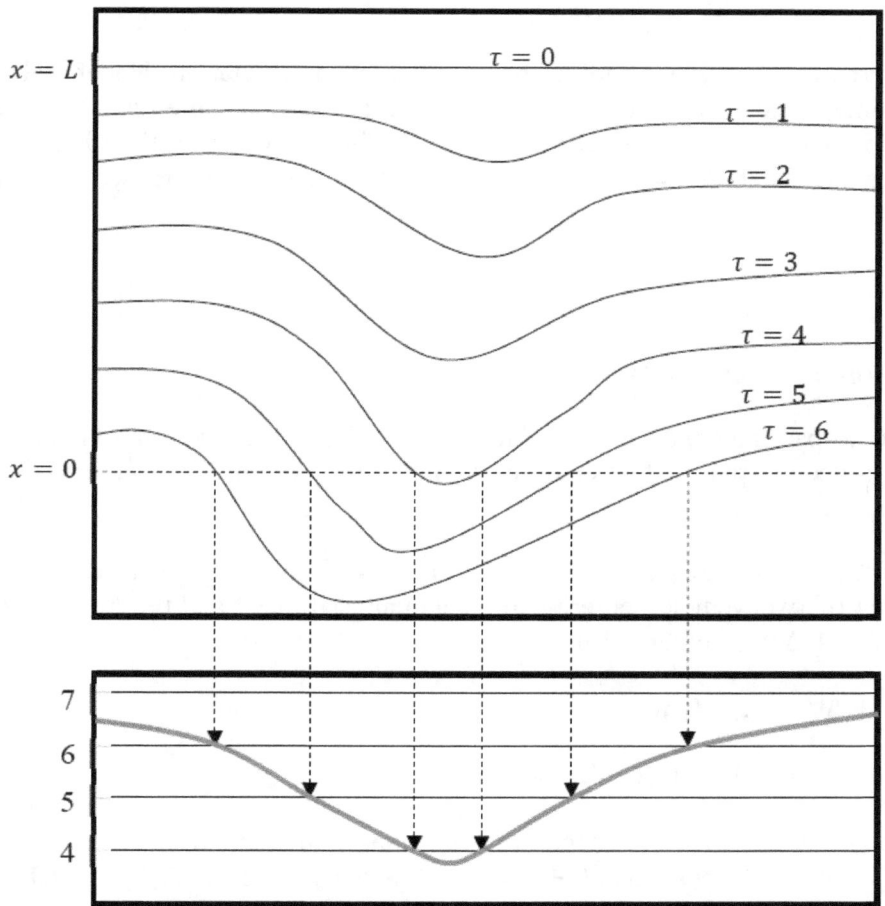

Figure 4. Concept of travel time as distance from downstream boundary in a wave propagation paradigm.

In the following sections some desirable properties of travel time such as stability and first-in first-out are discussed and relevant conditions on travel time derivatives are presented. Later, the space-time partial differential equation representing relationship between partial derivatives of travel time and local speed is derived and its properties in relations with the feasible region of local travel time variations are discussed. Also, two efficient finite difference schemes to solve for travel times given speeds are presented.

3.2.1 Travel Time Properties and Modeling

3.2.1.1 Stability

Under stable traffic flow condition, speeds over time and space do not change. In terms of travel times, stability in time direction suggests that travel time at a given point on the highway will be constant over time, or in other words

$$\tau_t = 0 \tag{24}$$

$$\tau_{tt} = 0 \tag{25}$$

In the space direction, however, stability has a different interpretation. Under stable conditions, at any given time, it takes a vehicle $\Delta t = \Delta x/v = n\Delta x$ additional time units to cover a distance Δx in front of it. This time lapse is effectively the difference in travel time between two points along the space coordinate $\Delta \tau = \tau(x + \Delta x, .) - \tau(x, .) = -\Delta t$. Therefore, we can write

$$\tau_x = -1/v = -n \tag{26}$$

$$\tau_{xx} = 0 \tag{27}$$

3.2.1.2 First-In-First-Out (FIFO)

To enforce the no-passing condition on vehicle trajectories in the solution domain, it is necessary to require any trajectory pair that entered the segment in a given order to leave the segment in the same order and vice versa.

It should be noted that the forward FIFO condition is well-known and can be derived by forcing departure times of two vehicles entering the segment at times t and t + Δt to follow the same order $D(t) \leq D(t + \Delta t)$. In other words,

$$t + \tau(., t) \leq t + \Delta t + \tau(., t + \Delta t) \tag{28}$$

$$\tau_t = \lim_{\Delta t \to 0} (\tau(., t + \Delta t) - \tau(., t))/\Delta t \geq -1 \tag{29}$$

However, the backward FIFO is to enforce the condition that arrival times of two vehicles departing the segment at times θ and $\theta + \Delta\theta$ to follow the same order $A(\theta) \leq A(\theta + \Delta\theta)$

$$\theta - \delta(., \theta) \leq \theta + \Delta\theta - \delta(., \theta + \Delta\theta) \tag{30}$$

$$\delta_\theta = \lim_{\Delta\theta \to 0} (\delta(., \theta + \Delta\theta) - \delta(., \theta))/\Delta\theta \leq 1 \tag{31}$$

It should be noted that the realized travel time of a vehicle departing the segment at time θ, $\delta(., \theta)$, is in fact the same as the travel time of the vehicle when it arrived in the segment at time t, $\tau(., t)$. Therefore, we can deduce that

$$\tau_t = \delta_\theta \leq 1 \tag{32}$$

Similarly, in the space direction we can write

$$\tau(x, .) \geq \Delta x/v(x, .) + \tau(x + \Delta x, .) \tag{33}$$

$$\tau_x = \lim_{\Delta x \to 0} (\tau(x + \Delta x, .) - \tau(x, .))/\Delta x \leq -1/v(x, .) = -n \tag{34}$$

3.2.2 First Order Travel Time Model and Finite Difference Solution Schemes

Let $\tau(x, t)$ represent travel time from a point (x, t) in space x and time t coordinates to a given downstream point x_d. This definition specifies the so called a priori travel time since at point (x, t) travel time τ(x, t) has not yet realized. It should be noted that in what follows derivations and proposed solution schemes are based on this definition of travel time. However, it will be trivial to derive similar models in the case of a posteriori travel times.

Assuming smoothness and therefore existence of derivatives we can use Taylor's function expansion to obtain the travel time near a point (x, t) as

$$\tau(x + dx, t + dt) = \tau(x, t) + \tau_t dt + \tau_x dx + O((dt)^2) + O((dx)^2) \tag{35}$$

Figure 5 illustrates these definitions and the above relationship where the pair of points (x, t) and (x + dx, t + dt) are located on a single vehicle trajectory. In this case, it is obvious that travel time to downstream at the second point has a simple relationship with the travel time at the first point, or more specifically we can write

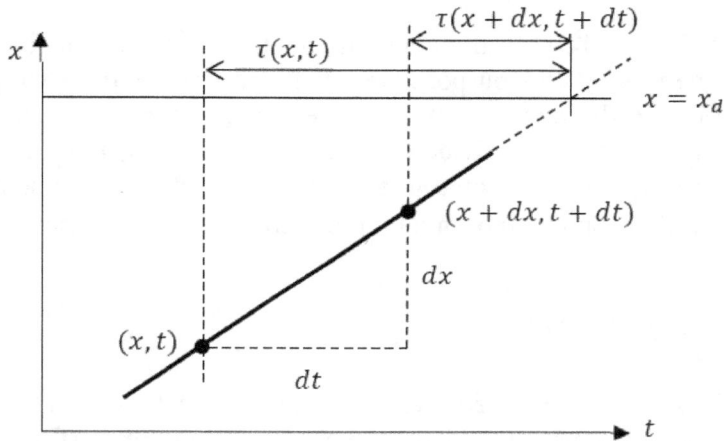

Figure 5. Schematic illustration of a vehicle trajectory in space-time domain

$$\tau(x + dx, t + dt) = \tau(x, t) - dt \tag{36}$$

Substituting (36) into (35), τ(x, t) term cancels out on both sides and then moving derivative terms to the right hand side we obtain

$$\tau_t dt + \tau_x dx = -dt + O((dt)^2) + O((dx)^2) \tag{37}$$

Now dividing both sides of (37) by dt we get

$$\tau_t + \tau_x \left(\frac{dx}{dt}\right) = -1 + O(dt) + O\left(\frac{(dx)^2}{dt}\right) \tag{38}$$

As dt goes to zero we know in the limit $\frac{dx}{dt}$ goes to speed v(x, t) which in vehicular traffic is typically assumed to be bounded from above by a known free flow speed v_f. Therefore, taking the limits from both sides of (38) we obtain the following equation

$$\tau_t + v\tau_x = -1 \tag{39}$$

which is a first order linear PDE that represents the relationships between partial derivatives of travel time in space and time directions at any given point. At a given location when τ_t is positive it means that travel times are increasing as time goes by, while τ_t negative suggests that travel times at that location are decreasing. At steady state ($\tau_t = 0$) travel time does not change over time. To enforce FIFO condition it is necessary that $-1 \leq \tau_t \leq 1$. Thus according to (39) in the extremes of this bound the travel time derivative in the space direction will be defined by

$$\tau_x = \begin{cases} \frac{-2}{v} & \text{if } \tau_t = 1, v > 0 \\ \frac{-1}{v} & \text{if } \tau_t = 0, v > 0 \end{cases} \tag{40a}$$

$$v\tau_x = 0 \quad \text{if } \tau_t = -1 \tag{40b}$$

Travel time model (39) can be solved numerically using a forward-time backward-space (FTBS) finite difference scheme. For this purpose we need to discretize the solution domain into cells. First, time duration T is divided into N time intervals $\{t_n | n = 0,1,\ldots,N\}$ each of length $\Delta t = T/N$ and the highway length X is divided into M space cells $\{x_i | i = 0,1,\ldots,M\}$ each of length $\Delta x = X/M$. To each space cell x_i at time interval t_n, a discrete average speed U_n^i and travel time θ_n^i is assigned. Therefore, under smooth conditions travel time evolution at each space cell over time is given by,

$$\theta_{n+1}^i = \theta_n^i - \frac{\Delta t}{\Delta x} U_{n+1}^i (\theta_n^i - \theta_n^{i-1}) - \Delta t \tag{41}$$

It should be noted that this scheme requires an estimate of speed U_{n+1}^i at each update. The speed estimates can be obtained from a Godunov finite difference scheme (CTM-v) proposed to solve LWR-v model such as (6), (7), (11) and (12) which is summarized here as following

$$v_t + R(v)_x = 0 \tag{42}$$

$$U_{n+1}^i = U_n^i - \frac{\Delta t}{\Delta x} \left(g(U_n^i, U_n^{i+1}) - g(U_n^{i-1}, U_n^i) \right) \tag{43}$$

where the flow g is defined as,

$$g(v_1, v_2) = \begin{cases} R(v_2) & \text{if } v_1 \leq v_2 \leq v_c \\ R(v_c) & \text{if } v_1 \leq v_c \leq v_2 \\ R(v_1) & \text{if } v_c \leq v_1 \leq v_2 \\ \max(R(v_1), R(v_2)) & \text{if } v_1 \geq v_2 \end{cases} \tag{44}$$

where v_c is the minimum of the convex flux function. In the case of Greenshield's flux (45), we will have $v_c = \frac{v_f}{2}$.

$$R(v) = v^2 - v_f v \tag{45}$$

Also, in this case, to ensure stability of the numerical discretization method, the length of both spatial and time steps must meet the Courant-Friedrichs-Lewy (CFL) condition:

$$\left|\frac{\Delta t}{\Delta x}\max(R'(v))\right| \leq 1 \tag{46}$$

The CFL condition suggests that no entering vehicle at the beginning of a time interval can exit the cell in that same time interval. Thus, in the case of Greenshield's flux function, this condition translates to

$$v_f \Delta t \leq \Delta x \tag{47}$$

It should be noted that scheme (41) is expected to work well under smooth conditions. However, this assumption does not typically hold under real traffic conditions where abrupt jumps (shockwaves) in traffic variables such as density, speed and therefore travel times are prevalent. In the next section, we derive a conservative form of the travel time model which renders itself to more robust finite difference schemes in presence of discontinuities.

3.2.3 Conservative Travel Time Model

To derive the conservative form of the travel time model let's consider the first order LWR-v velocity based continuum traffic flow model (42) to represent the speed evolutions in conjunction with the first order linear travel time model (39) presented earlier.

$$\begin{cases} v_t + R(v)_x = 0 \\ \tau_t + v\tau_x = -1 \end{cases} \tag{48}$$

Assuming free flow speed is constant over time and by multiplying LWR-v and travel time models by τ and $(v - v_f)$ respectively we obtain the following equivalent statements for both models

$$\begin{cases} \tau(v - v_f)_t + \tau R(v)_x = 0 \\ (v - v_f)\tau_t + v(v - v_f)\tau_x = v_f - v \end{cases} \tag{49}$$

Recalling (45) the Greenshields' LWR-v flux function, we can sum up the two sides of the above pair of equations to obtain

$$[(v - v_f)\tau]_t + [v(v - v_f)\tau]_x = v_f - v \tag{50}$$

Setting $(v - v_f)\tau = u$ in the above equation we arrive at the standard conservative form of the travel time model

$$u_t + (vu)_x = v_f - v \tag{51}$$

This is a nonlinear nonhomogeneous hyperbolic PDE which is a conservation law with linear source term. The conserved variable u can be interpreted as the excess distance a vehicle would have covered during its travel time τ had it traveled at free flow speed v_f all the way instead of its spot speed v. It should be noted that in solving for travel time τ it is preferred to work with (50) due to its explicit form. These solutions are designed to take into account the fact that despite smoothness assumptions discontinuities (shockwaves) are present in the solution domain.

To obtain weak solutions of this PDE it is possible to use any standard finite difference scheme which satisfies the entropy conditions. Here we propose a Godunov like finite difference scheme such as the following to solve this model.

$$\theta_{n+1}^i = \frac{(U_h^i - v_f)\theta_h^i - \frac{\Delta t}{\Delta x} U_{n+1}^i \left(F(U_h^i, U_h^{i+1}, \theta_h^i, \theta_h^{i+1}) - F(U_h^{i-1}, U_h^i, \theta_h^{i-1}, \theta_h^i) \right) + \Delta t \left(v_f - \frac{U_h^i + U_{n+1}^i}{2} \right)}{U_{n+1}^i - v_f} \quad (52)$$

where,

$$F(v_1, v_2, \tau_1, \tau_2) = f(v^*, \tau^*) = v^*(v^* - v_f)\tau^* = g(v_1, v_2)\tau^* \quad (53)$$

and due to the fact that Greenshields velocity flux R(v) is non-positive and FIFO conditions ($\tau_1 \geq \tau_2$), we define

$$\tau^* = \begin{cases} \tau_1 & \text{if } (v_1 - v_f)\tau_1 \leq (v_2 - v_f)\tau_2 \\ \tau_2 & \text{if } (v_1 - v_f)\tau_1 > (v_2 - v_f)\tau_2 \end{cases} \quad (54)$$

It should be noted that here similar to previous scheme we can use CTM-v model given by (6), (7), (11) and (12) to estimate speeds used in travel time estimation.

In our application, it should be noted that boundary values are very easy to determine. In fact, at downstream point the value of travel time function is constantly equal to zero

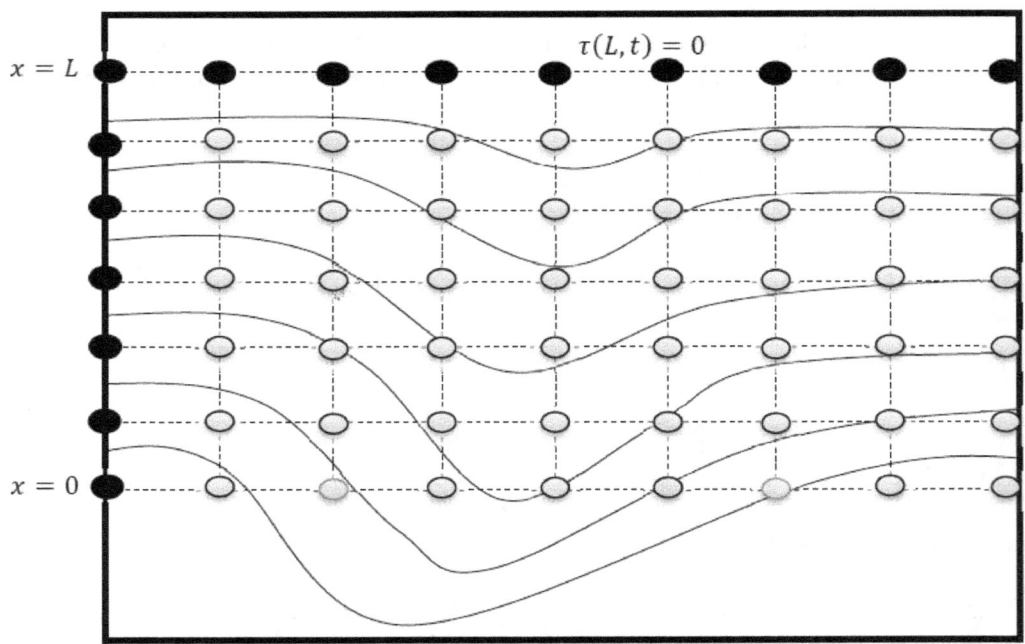

Figure 6. Space-time grid representation of the solution domain.

In our application, it should be noted that boundary values are very easy to determine. In fact, at downstream point the value of travel time function is constantly equal to zero

$$\tau(L, t) = 0 \tag{55}$$

Initial conditions are also assumed to be known at every point along the segment under consideration

$$\tau(x, 0) = F(x) \tag{56}$$

Travel time data provided by AVI technologies such as Bluetooth detection units can serve as additional boundary, initial or internal conditions. In our specific application such data sources are considered as boundary conditions since the pair of detectors are assumed to be placed at both ends of the segment of interest

$$\tau(0, t_i) = G(t_i) \tag{57}$$

Figure 6 shows a typical grid in which initial and downstream boundary conditions on travel times are represented by dark nodes, while red nodes represent occasional travel time measurements at upstream boundary.

3.3 State Space Models

First, a brief overview of dynamical systems is given. The most general representation of a dynamical system can be given as

$$\begin{cases} \dot{x} = f(x, u, w) \\ y = h(x, v) \end{cases} \tag{58}$$

where,

x, is the system state

u, is the system input

w, is the system process noise

y, is the measurements obtained from the system

v, is the measurement noise

f, is the system dynamic function, and

h, is the measurement function.

If $f(.)$ and $h(.)$ are explicit functions of time t then the system is time-varying. Otherwise, the system is time invariant. Also, it should be noted that if $f(.)$ and $h(.)$ are linear functions then the system is linear,

$$\begin{cases} \dot{x} = Ax + Bu + w \\ y = Hx + v \end{cases} \quad (59)$$

where,

A, is the system matrix,

B, is the input matrix, and

H, is the output matrix.

State estimation under the general model given above is possible yet it is not easy in most practical cases. In fact, two problems need to be resolved with regard to the general state-space model:

- Nonlinearity

- Continuity

In this study, we propose the following two solutions to deal with each problem:

- Linearization

- Discretization

In the following sections basics of each step in deriving a linear and discrete state space model are presented.

3.3.1 Linearization

To linearize the model the right hand side of the system and measurement equations need to be expanded around a given point using Taylor's expansion

$$\begin{cases} \dot{x} = f(\bar{x}, \bar{u}, \bar{w}) + \left.\frac{\partial f}{\partial x}\right|_{\bar{x}} (x - \bar{x}) + \left.\frac{\partial f}{\partial u}\right|_{\bar{u}} (u - \bar{u}) + \left.\frac{\partial f}{\partial w}\right|_{\bar{w}} (w - \bar{w}) \\ y = h(\bar{x}, \bar{v}) + \left.\frac{\partial h}{\partial x}\right|_{\bar{x}} (x - \bar{x}) + \left.\frac{\partial h}{\partial v}\right|_{\bar{v}} (v - \bar{v}) \end{cases} \quad (60)$$

Noting that

$$\dot{\bar{x}} = f(\bar{x}, \bar{u}, \bar{w}) \quad (61)$$

$$\bar{y} = h(\bar{x}, \bar{v}) \quad (62)$$

we can write

$$\begin{cases} \dot{x} - \dot{\bar{x}} = \frac{\partial f}{\partial x}\Big|_{\bar{x}} (x - \bar{x}) + \frac{\partial f}{\partial u}\Big|_{\bar{u}} (u - \bar{u}) + \frac{\partial f}{\partial w}\Big|_{\bar{w}} (w - \bar{w}) \\ y - \bar{y} = \frac{\partial h}{\partial x}\Big|_{\bar{x}} (x - \bar{x}) + \frac{\partial h}{\partial v}\Big|_{\bar{v}} (v - \bar{v}) \end{cases} \quad (63)$$

And, setting the differences as new variables,

$$\tilde{x} = x - \bar{x} \quad (64)$$

$$\tilde{u} = u - \bar{u} \quad (65)$$

$$\tilde{w} = w - \bar{w} \quad (66)$$

$$\tilde{v} = v - \bar{v} \quad (67)$$

we can write

$$\begin{cases} \dot{\tilde{x}} = \frac{\partial f}{\partial x}\Big|_{\bar{x}} \tilde{x} + \frac{\partial f}{\partial u}\Big|_{\bar{u}} \tilde{u} + \frac{\partial f}{\partial w}\Big|_{\bar{w}} \tilde{w} \\ \tilde{y} = \frac{\partial h}{\partial x}\Big|_{\bar{x}} \tilde{x} + \frac{\partial h}{\partial v}\Big|_{\bar{v}} \tilde{v} \end{cases} \quad (68)$$

or, simply

$$\begin{cases} \dot{\tilde{x}} = A\tilde{x} + B\tilde{u} + C\tilde{w} \\ \tilde{y} = H\tilde{x} + D\tilde{v} \end{cases} \quad (69)$$

where, definitions of matrices A, B, C, H, and D are self-evident.

3.3.2 Discretization

Transformation from continuous time to discrete time dynamics is presented below.

$$\tilde{x}(t) = e^{A(t-t_0)}\tilde{x}(t_0) + \int_{t_0}^{t} e^{A(t-\tau)} B\tilde{u}(\tau)d\tau + \int_{t_0}^{t} e^{A(t-\tau)} C\tilde{w}(\tau)d\tau \quad (70)$$

Let $t = t_k$ (some discrete time point) and let the initial time $t_0 = t_{k-1}$ (the previous discrete time point). Assume that $A(\tau)$, $B(\tau)$, $C(\tau)$, $\tilde{u}(\tau)$, and $\tilde{w}(\tau)$ are approximately constant in the interval of integration. We then obtain

$$\tilde{x}(t_k) = e^{A(t_k - t_{k-1})}\tilde{x}(t_{k-1}) + \int_{t_{k-1}}^{t_k} e^{A(t_k-\tau)} d\tau\, B\tilde{u}(t_{k-1}) + \int_{t_{k-1}}^{t_k} e^{A(t_k-\tau)} d\tau\, C\tilde{w}(t_{k-1}) \quad (71)$$

Now let's define $\Delta t = t_k - t_{k-1}$, and $\alpha = \tau - t_{k-1}$, and substitute for τ in the above equation to obtain

$$\tilde{x}(t_k) = e^{A\Delta t}\tilde{x}(t_{k-1}) + e^{A\Delta t} \int_0^{\Delta t} e^{-A\alpha} d\alpha\, B\tilde{u}(t_{k-1}) + e^{A\Delta t} \int_0^{\Delta t} e^{-A\alpha} d\alpha\, C\tilde{w}(t_{k-1}) \quad (72)$$

To further simplify please note that if A is invertible (nonsingular), then we can write

$$\int_0^{\Delta t} e^{-A\alpha} d\alpha = [I - e^{-A\Delta t}]A^{-1} \tag{73}$$

And therefore, the linear discrete time system dynamic equation can be summarized as

$$\tilde{x}(t_k) = F\tilde{x}(t_{k-1}) + G\tilde{u}(t_{k-1}) + L\tilde{w}(t_{k-1}) \tag{74}$$

where,

$$F = e^{A\Delta t} \tag{75}$$

$$G = F[I - e^{-A\Delta t}]A^{-1}B \tag{76}$$

$$L = F[I - e^{-A\Delta t}]A^{-1}C \tag{77}$$

But, when A is not invertible (singular) then computing the matrix exponential integral is not as straight-forward. In this case there are two approaches to this problem. The first approach consists of employing interpolating polynomial. This method besides giving an approximation is not conducive to straight-forward implementation. The second approach, namely choosing a sub-matrix of an extended matrix exponential is adopted in this study. We define,

$$\begin{cases} \dot{\tilde{X}} = \tilde{A}\tilde{X} \\ \tilde{y} = \tilde{H}\tilde{X} + D\tilde{v} \end{cases} \tag{78}$$

where,

$$\tilde{X} = \begin{bmatrix} \tilde{x} \\ \tilde{u} \\ \tilde{w} \end{bmatrix} \tag{79}$$

$$\tilde{A} = \begin{bmatrix} A & B & C \\ 0 & 0 & 0 \\ 0 & 0 & 0 \end{bmatrix} \tag{80}$$

$$\tilde{H} = [H \quad 0 \quad 0] \tag{81}$$

Now, this can be expressed easily in the discretized form,

$$\tilde{X}(t_k) = \tilde{F}\tilde{X}(t_{k-1}) \tag{82}$$

where,

$$\tilde{F} = e^{\tilde{A}\Delta t} \tag{83}$$

$$\tilde{F} = \begin{bmatrix} F & G & L \\ - & - & - \\ - & - & - \end{bmatrix} \tag{84}$$

The generalized form is obtained by adding an error term to the system dynamics equation

$$\widetilde{X}(t_k) = \widetilde{F}\widetilde{X}(t_{k-1}) + \widetilde{W}_{k-1} \tag{85}$$

$$\widetilde{W}_k = \begin{bmatrix} 0 \\ 0 \\ \xi_k \end{bmatrix} \tag{86}$$

$$\xi_k \sim N(0, \sigma^2) \tag{87}$$

It should be noted that essential properties of a dynamical system from an analytical perspective include the following:

- Stability

- Controllability and observability

- Stabilizability and detectability

These properties can be investigated using definitions and theorems given in standard textbooks on state space modeling and estimation (Simon, 2006).

While various methods are proposed to estimate system state under the most general state space models, the linear models are most useful in terms of their applicability and rigorous theoretical derivations. The traffic and travel time dynamical models given in 4.1 and 4.2 need to be recast in a state-space form. In this dynamical system, speeds and travel times are considered as system state. No inputs to the system are considered as our modeling effort does not include any traffic control measures. A subset of system state vector (speeds and travel times) is considered as measurement vector.

$$X_k = A_k X_{k-1} + B_k U_k + v_k \tag{88}$$

3.4 Optimal Estimation Methods

In this section estimation methods based on linear and nonlinear filtering approaches are introduced. Kalman filtering is a recursive root mean square error estimator that is optimal for linear state-space models with white noise errors. In case of nonlinear models and or correlated non-Gaussian errors, Kalman filtering is still applicable but only will be the best linear estimator. In such cases, extensions of Kalman filtering such as Extended Kalman filtering (EKF), Unscented Kalman filtering (UKF) and Particle filtering (PF) are more appropriate. It should be noted that these are least square methods which attempt to minimize the overall variance of the error by propagating the mean and variance of the estimation errors. Using the notation of the norm in Hilbert spaces, Kalman filter and its extensions minimize the H_2 norm of the system state estimates.

More recently based on a game theory concept methods to minimize the maximum estimation error are introduced which attempt to minimize the H_∞ norm of the Hilbert space. The H_∞

filtering is capable of dealing with non-Gaussian and correlated (non-white) errors in the estimation process. This is the main advantage of H_∞ filtering over H_2 (Kalman) filtering. However, the issue of dealing with nonlinear system and measurement equations still needs further attention. In this segment nonlinear H_∞ filters analogous to extensions of Kalman filtering are introduced.

In practice we may experience delayed arrival of measurements at the processing unit or we may miss or rule out some measurement altogether in the estimation process. Also, and especially in the case of travel time measurement there is a high chance of out of order measurement arrivals. In this segment, remedies and methods to account for these effects effectively and efficiently in the estimation process are introduced.

In practice, out of sequence arrival of measurements are common place in multi-sensor central processing systems. Essentially, two main approaches to accounting for delayed measurements in the estimation process exist. The first approach is comprised of step by step update of states from time to which the delayed measurement belongs to the current time. This would virtually require a parallel update process. Instead, the second method is essentially trying to update the current state in a giant leap. The two methods have different implications in terms of their efficiency, processing and storage requirements.

3.4.1 DISCRETE KALMAN FILTER

Considering a linear discrete-time dynamic system,

$$v_{n+1} = M_n v_n + w_n \tag{89}$$

with measurements, y_n, that are linearly related to the state of the system at each time interval:

$$y_n = H_n v_n + u_n \tag{90}$$

and, the set of additional assumptions that system and measurement errors are white and not correlated:

$$E(w_n w_k^T) = Q_n \delta_{n-k} \tag{91}$$

$$E(u_n u_k^T) = R_n \delta_{n-k} \tag{92}$$

$$E(w_n u_k^T) = 0, \qquad \forall n, k \tag{93}$$

It is desired to find the best estimate of the system at each time interval in a least square sense. Kalman filter is a linear recursive estimator of the a posteriori state of the system, $\hat{v}_n^+$ based on the a priori estimate of the state, $\hat{v}_n^-$, and the new measurement y_n:

$$\hat{v}_n^+ = \hat{v}_n^- + K_n(y_n - H_n \hat{v}_n^-) \tag{94}$$

where,

K_n is the estimator gain matrix that minimizes the sum of the variances of all estimation errors,

$\tilde{e}_n = v_n - \hat{v}_n$, to be optimally determined by solving the optimization problem:

$$\min E[\tilde{e}_n^T \tilde{e}_n] \tag{95}$$

Subject to: (22-27)

The Kalman filter is initialized as follows:

$$\hat{v}_0^+ = E(v_0) \tag{96}$$

$$P_0^+ = E[(v_0 - \hat{v}_0^+)(v_0 - \hat{v}_0^+)^T] \tag{97}$$

Then, at each iteration, a priori and a posteriori estimates of the state and their corresponding covariance matrices are computed using the following equations:

$$P_n^- = F_{n-1} P_{n-1}^+ F_{n-1}^T + Q_{n-1} \tag{98}$$

$$K_n = P_n^- H_n^T (H_n P_n^- H_n^T + R_n)^{-1}$$
$$= P_n^+ H_n^T R_n^{-1} \tag{99}$$

$$\hat{v}_n^- = M_{n-1} \hat{v}_{n-1}^+ \tag{100}$$

$$\hat{v}_n^+ = \hat{v}_n^- + K_n(y_n - H_n \hat{v}_n^-) \tag{101}$$

$$P_n^+ = (I - K_n H_n) P_n^- (I - K_n H_n)^T + K_n R_n K_n^T$$
$$= [(P_n^-)^{-1} + H_n^T R_n^{-1} H_n]^{-1}$$
$$= (I - K_n H_n) P_n^- \tag{102}$$

3.4.2 DELAYED MEASUREMENT KALMAN FILTER

The standard Kalman filter is designed to make optimal estimates of the state variables on the basis of measurements that arrive at the processing unit in a sequential order. In other words, at current time interval n in the Kalman filter, we have a priori state and covariance estimates $\hat{v}_n^-$, and P_n^- that take into account measurements up to and including time interval $n-1$. Also, the a posteriori state and covariance estimates, $\hat{v}_n^+$, and P_n^+ based on measurements up to and including time interval n are assumed to be known. In practice, however, it is common place to receive measurement(s) v_{n_0}, where $n_0 < n$. To account for these delayed measurements then we should update the state estimate and its covariance to $\hat{v}_{nn_0}$, and P_{nn_0}, respectively. The strategy normally adopted to tackle this problem is first to retrospectively predict (retrodict) the state estimate from time interval n back to interval n_0,

$$S_n = H_n P_n^- H_n^T + R_n \tag{103}$$

$$\hat{v}_{n_0 n} = F_{n_0 n}[\hat{v}_n - Q_{nn_0} H_n^T S_n^{-1} r_n] \tag{104}$$

where,

$$r_n = y_n - H_n \hat{v}_n^-, \text{ and} \tag{105}$$

$$S_n = Cov(r_n) \tag{106}$$

Then we compute the covariance of the retrodicted state using the following equations

$$P_w(n, n_0) = Q(n, n_0) - Q(n, n_0)H^T(n)S^{-1}(n)H(n)Q(n, n_0) \tag{107}$$

$$P_{vw}(n, n_0) = Q(n, n_0) - P^{-1}(n)H^T(n)S^{-1}(n)H(n)Q(n, n_0) \tag{108}$$

$$P(n_0, n) = F(n_0, n)\{P(n) - P_{vw}(n, n_0) - P_{vw}^T(n, n_0) + P_w(n, n_0)\}F^T(n_0, n) \tag{109}$$

Now we can compute the covariance of the retrodicted measurement at time n_0 using

$$S(n_0) = H(n_0)P(n_0, n)H^T(n_0) + R(n_0) \tag{110}$$

Then we can compute the covariance of the state at time n and the retrodicted measurement at time n_0 using the following

$$P_{vy}(n, n_0) = [P(n) - P_{vw}(n, n_0)]F^T(n_0, n)H^T(n_0) \tag{111}$$

Finally, we use the delayed measurement $y(n_0)$ to update the state estimate and its covariance

$$\hat{v}(n, n_0) = \hat{v}(n) + P_{vy}(n, n_0)S^{-1}(n_0)[y(n_0) - H(n_0)\hat{v}(n_0, n)] \tag{112}$$

$$P(n, n_0) = P(n) + P_{vy}(n, n_0)S^{-1}(n_0)P_{vy}^T(n, n_0) \tag{113}$$

It should be noted that the computational cost of this delayed measurement filter can be reduced by considering some simplifying approximations with slight reduction in the accuracy of the estimates.

3.4.3 H_∞ FILTER

As before, consider the system equations (84), (85) along with the linear transformation of state variable v_n,

$$z_n = L_n v_n \tag{114}$$

Please note that we are not making any assumptions about system model and estimation errors w_n, and u_n. Our goal is to estimate z_n. It is obvious that if we let transformation matrix L_n to be equal to the identity matrix I, then $z_n = v_n$.

The cost function in the estimation problem is given as

$$J_1 = \frac{\sum_{n=0}^{N-1} \|z_n - \hat{z}_n\|_{S_k}^2}{\|x_0 - \hat{x}_0\|_{P_0^{-1}}^2 + \sum_{n=0}^{N-1}\left(\|w_n\|_{Q_n^{-1}}^2 + \|v_n\|_{R_n^{-1}}^2\right)} \tag{115}$$

where, P_0, Q_n, R_n and S_n are symmetric, positive definite matrices chosen by the user for the specific problem.

The cost function (110) can be made to be less than a user specified threshold $1/\theta$ with the following estimation strategy,

$$\bar{S}_n = L_n^T S_n L_n \tag{116}$$

$$K_n = P_n[I - \theta \bar{S}_n P_n + H_n^T R_n^{-1} H_n P_n]^{-1} H_n^T R_n^{-1} \tag{117}$$

$$\hat{v}_{n+1} = M_n[\hat{v}_n + K_n(y_n - H_n \hat{v}_n)] \tag{118}$$

$$P_{n+1} = M_n P_n[I - \theta \bar{S}_n P_n + H_n^T R_n^{-1} H_n P_n]^{-1} M_n^T + Q_n \tag{119}$$

However, the above solution is valid only if the following necessary condition holds at each time step

$$P_n^{-1} - \theta \bar{S}_n + H_n^T R_n^{-1} H_n > 0 \tag{120}$$

3.4.4 MIXED KALMAN/H_∞ FILTER

We would like to have a filter that combines the best features of Kalman filtering with the best features of H_∞ filtering.

Steady-state Kalman filter cost function is

$$J_2 = \lim_{N \to \infty} \sum_{n=0}^{N} E(\|v_n - \hat{v}_n\|_2) \tag{121}$$

Let S_k and L_k are identity matrices, then the steady-state H_∞ estimator cost function is

$$J_\infty = \lim_{N \to \infty} \max_{x_0, w_n, u_n} \frac{\sum_{n=0}^{N} \|v_n - \hat{v}_n\|^2}{\|x_0 - \hat{x}_0\|_{P_0^{-1}}^2 + \sum_{n=0}^{N} \left(\|w_n\|_{Q_n^{-1}}^2 + \|u_n\|_{R_n^{-1}}^2\right)} \tag{122}$$

Given an n-state observable LTI system

$$v_{n+1} = M \ v_n + w_n \tag{123}$$

$$y_n = H \ v_n + u_n \tag{124}$$

Find an estimator of the form

$$\hat{v}_{n+1} = \widehat{M} \ v_n + K y_n \tag{125}$$

Step 1: Find the n*n positive semidefinite matrix P that satisfies the following Riccati equation

$$P = MPM^T + Q + MP(I/\theta^2 - P)^{-1} PM^T - P_a V^{-1} P_a^T \tag{126}$$

where P_a and V are defined as

$$P_a = MPH^T + MP(I/\theta^2 - P)^{-1}PH^T \tag{127}$$

$$V = R + HPH^T + HP(I/\theta^2 - P)^{-1}PH^T \tag{128}$$

Step 2: Derive the $\widehat{M}$ and K matrices as

$$K = P_a V^{-1} \tag{129}$$

$$\widehat{M} = M - KH \tag{130}$$

The estimator of equation (120) satisfies the mixed Kalman/H_∞ estimation problem if and only if $\widehat{M}$ is stable. In this case, the state estimation error satisfies the bound

$$\lim_{n \to \infty} \sum_{n=0}^{N} \|v_n - \hat{v}_n\|^2 \leq Tr(P) \tag{131}$$

3.4.5 ROBUST KALMAN/H_∞ FILTER

$$v_{n+1} = (M_k + \Delta M_k)v_n + w_n \tag{132}$$

$$y_n = (H_k + \Delta H_k)v_n + u_n \tag{133}$$

$$\begin{bmatrix} \Delta M_n \\ \Delta H_n \end{bmatrix} = \begin{bmatrix} M_{1n} \\ M_{2n} \end{bmatrix} \Gamma_n N_n \tag{134}$$

where, M_{1n}, M_{2n}, and N_n are known matrices, and Γ_n is an unknown matrix for which ($\Gamma_n^T \Gamma_n - I$) is a negative semidefinite matrix.

The problem is to define a state estimator of the form

$$\hat{v}_{n+1} = \widehat{M}_n \hat{v}_n + K_n y_n \tag{135}$$

with the following properties:

1. The estimator is stable (i.e., the eigenvalues of $\widehat{M}_n$ are less than one in magnitude).

2. The estimation error $\tilde{e}_n$ satisfies the following worst-case bound

$$\max_{w_n, u_n} \frac{\|\tilde{e}_n\|^2}{\|w_n\|^2 + \|u_n\|^2 + \|\tilde{e}_0\|_{S_1^{-1}}^2 + \|v_0\|_{S_2^{-1}}^2} < \frac{1}{\theta} \tag{136}$$

3. The estimation error $\tilde{e}_n$ satisfies the following RMS bound

$$E[\tilde{e}_n^T \tilde{e}_n] < P_n \tag{137}$$

The solution to this problem can be found by the following procedure

Step 1: Choose some scalar sequence $\alpha_n > 0$, and a small scalar $\epsilon > 0$.

Step 2: Define the following matrices:

$$R_{11n} = Q_n + \alpha_n M_{1n} M_{1n}^T \tag{138}$$

$$R_{12n} = \alpha_n M_{1n} M_{2n}^T \tag{139}$$

$$R_{22n} = R_n + \alpha_n M_{2n} M_{2n}^T \tag{140}$$

<u>Step 3:</u> Initialize P_n and $\tilde{P}_n$ as follows:

$$P_0 = S_1 \tag{141}$$

$$\tilde{P}_0 = S_2 \tag{142}$$

<u>Step 4:</u> Find positive definite solutions P_n and $\tilde{P}_n$ satisfying the following Riccati equations:

$$P_{n+1} = M_{1n} T_n M_{1n}^T + R_{11n} + R_{11n} R_{2n} R_{11n}^T -$$
$$[M_{1n} T_n H_{1n}^T + R_{11n} R_{2n} R_{12n} + R_{12n}] R_n^{-1} [\ldots]^T + \epsilon I \tag{143}$$

$$\tilde{P}_{n+1} = M_n \tilde{P}_n M_n^T + M_n \tilde{P}_n N_n^T (\alpha_n I - N_n \tilde{P}_n N_n^T)^{-1} N_n \tilde{P}_n M_n^T + R_{11n} + \epsilon I \tag{144}$$

where the matrices $R_{1n}, R_{2n}, M_{1n}, H_{1n}$ and T_n are defined as

$$R_{1n} = (\tilde{P}_n^{-1} - N_n^T N_n / \alpha_n)^{-1} M_n^T \tag{145}$$

$$R_{2n} = R_{1n}^{-1} (\tilde{P}_n^{-1} - N_n^T N_n / \alpha_n)^{-1} R_{1n}^{-T} \tag{146}$$

$$M_{1n} = M_n + R_{11n} R_{1n}^{-1} \tag{147}$$

$$H_{1n} = H_n + R_{12n}^T R_{1n}^{-1} \tag{148}$$

$$T_n = (P_n^{-1} - \theta^2 I)^{-1} \tag{149}$$

<u>Step 5:</u> If the Riccati equation solutions satisfy

$$\frac{1}{\theta^2} I > P_n \tag{150}$$

$$\alpha_n I > N_n \tilde{P}_n N_n^T \tag{151}$$

then the estimator of equation ( ) solves the problem with

$$\tilde{R}_n = H_{1n} T_n H_{1n}^T + R_{12n}^T R_{2n} R_{12n} + R_{22n} \tag{152}$$

$$K_n = [M_{1n} T_n H_{1n}^T + R_{11n} R_{2n} R_{12n} + R_{12n}] \tilde{R}_n^{-1} \tag{153}$$

$$\widehat{M}_n = M_{1n} - K_n H_{1n} \tag{154}$$

3.4.6 CONSTRAINED KALMAN FILTER

Typically, some extra information is available on relationships between state variables. To enforce these relationships usually additional constraints can be added to the state space model. Constrained Kalman Filter algorithm are developed to handle these situations. Over the years different approaches have been proposed to solve constrained Kalman Filters. Some notable approaches are the following:

- Model Reduction
- Perfect Measurements
- Projection Approaches
    - Maximum probability approach
    - Least squares approach
    - General projection approach
- A pdf Truncation Approach

In this study, the general projection approach is adopted to handle any likely situation where flow dynamics model and measurement equations are accompanied by any additional constraint on measured travel times and speeds. this problem can be written in the following general form:

$$\tilde{v} = argmin_{\tilde{v}}(\tilde{v} - \hat{v})^T W (\tilde{v} - \hat{v}) \text{ such that } D\tilde{v} = d \tag{155}$$

where W is any positive definite weighting matrix. The solution to the above problem is

$$\tilde{v} = \hat{v} - W^{-1}D^T(DW^{-1}D^T)^{-1}(D\hat{v} - d) \tag{156}$$

If we set $W = P^{-1}$, it results in the minimum variance filter which is similar to the maximum probability approach. Alternatively, if we set $W = I$ our solution will be similar to that of least squares approach.

3.4.7 CONSTRAINED H_∞ FILTER

The constrained H_∞ filter can be summarized as follows.

Given a linear system

$$v_{n+1} = M_n v_n + w_n \tag{157}$$

$$y_n = H_n v_n + u_n \tag{158}$$

$$D_n v_n = d_n \tag{159}$$

where the last equation above specifies equality constraints on the state. We assume that the constraints are normalized so $D_n D_n^T = I$. The covariance of w_n is equal to Q_n, but it might have a zero or a nonzero mean (i.e., it might contain a deterministic component). The covariance of u_n is the identity matrix.

Step 1: Initialize the filter as follows

$$\hat{v}_0 = 0 \tag{160}$$

$$P_0 = E(v_0 v_0^T) \tag{161}$$

Step 2: At each time step $n = 0, 1, ...$, do the following

Step 2.1: Choose the tuning parameter matrix G_n to weight the deterministic, biased component of the process noise. If $G_n = 0$ then we are assuming that the process noise is zero-mean and we get the Kalman filter performance. As G_n increases we are assuming that there is more of a deterministic, biased component to the process noise. This gives us better worst-case error performance but worse RMS error performance.

Step 2.2: Compute the next state estimate as follows

$$U_n = D_n^T D_n \tag{162}$$

$$\Sigma_n = (P_n H_n^T H_n - P_n G_n^T G_n + I)^{-1} P_n \tag{163}$$

$$P_{n+1} = (I - U_{n+1}) M_n \Sigma_n M_n^T (I - U_{n+1}) + Q_n \tag{164}$$

$$K_n = (I - U_{n+1}) M_n \Sigma_n H_n^T \tag{165}$$

$$\hat{v}_{n+1} = M_n \hat{v}_n + K_n (y_n - H_n \hat{v}_n) \tag{166}$$

Step 2.3: Verify that

$$(I - G_n P_n G_n^T) \geq 0 \tag{167}$$

If not then the filter is not valid.

3.4.8 H₂ (KALMAN) FILTER EXTENSIONS

Kalman Filter (KF) is developed based on two major assumptions that system dynamics model is linear and also modeling and measurement errors are normally and independently distributed. These are essentially very strong assumptions as most systems are both nonlinear and have non Gaussian and correlated errors. To deal with nonlinear systems such as the traffic modeling system proposed here we need to use Extended Kalman Filter (EKF) which accounts for nonlinearity in system modeling. Similarly, Unscented Kalman Filter (UKF) and Particle Filter (PF) are standard solutions to the non-white noise issue as mentioned previously.

3.4.9 OUT OF SEQUENCE AND MISSING MEASUREMENTS

In practice, out of sequence arrival of measurements are common place in multi-sensor central processing systems. Essentially, two main approaches to accounting for delayed measurements in the estimation process exist. The first approach is comprised of step by step update of states from time to which the delayed measurement belongs to the current time. This would virtually require a parallel update process. Instead, the second method is essentially trying to update the current state in a giant leap. The two methods have different implications in terms of their efficiency, processing and storage requirements.

3.5 Summary

In this chapter, first a well-known first order continuum traffic flow model is adopted to represent the dynamics of the system. An equivalent form of this model in terms of speed is derived. This model provides a theoretical framework to understand and analyze traffic processes on a variety of roadway facilities. Two approximate solution methods of this model are introduced. A finite difference and a finite element method for solving the velocity based equivalent of the first-order continuum traffic flow model numerically are proposed.

Second, some desirable properties of travel time which are adopted in this study are briefly presented. Enforcing these properties on the estimates defines a feasible solution region for travel time partial derivatives. Then, interpreting travel time as distance to a boundary in space-time domain we introduce a framework to relate integrated lagrangean (travel time) data and local speeds. Relevant derivations based on the first order kinematics principle are presented. Two efficient finite difference schemes to solve for travel times given speeds is introduced.

Finally, to derive the optimal estimates from the resulting state space model in presence of errors in modeling and measurements we propose optimal filtering approach. Kalman filtering (H_2), H_∞ and their extensions for nonlinear models and measurement equations are introduced. In particular, extended Kalman filtering (EKF), unscented Kalman filtering (UKF) and particle filtering (PF) and their H_∞ equivalents are discussed. Two alternative approaches to incorporate travel time data as either a nonlinear and implicit measurement equation or an additional implicit side constraint are introduced. Methods to address missing and out of sequence measurements are introduced.

4 Methodology of Dynamic Travel Time Prediction

Previous studies use various methods to predict travel times by locating similar traffic trends in historical data sets. However, a travel time sequence is typically used in previous studies, which is a vector including a sequence of travel time values across different time intervals. Unlike previous studies of travel time prediction, a new algorithm is developed in this research by using the spatial and temporal traffic status information to obtain candidate traffic data from the historical travel database. The extracted candidate traffic data represent historical similar traffic patterns, and the candidates are aggregated to compute future travel times.

4.1 The Dynamic Travel Time Prediction Framework

The proposed algorithm comprises three stages: identify current traffic status, select similar traffic patterns from historical data, and predict travel times. The framework of the three stages is demonstrated in Figure 7. The current traffic status is initially selected to represent the traffic status of all freeway sections from short-past to the current time interval. The traffic status in this case is a matrix across temporal and spatial axes. Thereafter, the historical traffic speed data with the same dimension to current traffic status is selected as a candidate. Based on the dissimilarity to the current speed matrix, several candidates are extracted to represent the historical recurrent traffic patterns that are similar to the current status. Finally, the subsequent dynamic travel times of those candidates are aggregated to represent the travel time distributions in the future.

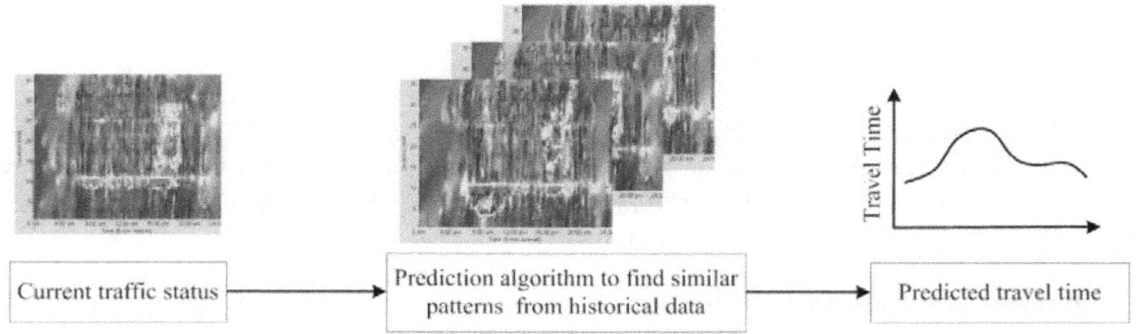

Figure 7. Framework of proposed dynamic travel time prediction algorithm.

4.2 Matching Traffic Patterns

A candidate selection scheme is proposed to select temporal-spatial traffic state candidates from a historical data set by matching with the real-time traffic state. Suppose c denotes the current time; the current traffic state $[c-L+1, c-L+2, ..., c]$ and the matching temporal-spatial traffic data $[t-L+1, t-L+2, ..., t]$ from a historical data set are denoted by tail time c and t, respectively. Here, L is the data length across time intervals to be matched. It should be noted that the traffic data of each time interval is a vector that covers all spatial sections (N sections) of the freeway stretch; therefore, the traffic data for L time intervals is a matrix with dimension L by N. Various template matching methods can be used to define the dissimilarity between the current traffic status and historical data, such as the Euclidean distance (Otsuka et al. 2000b, Otsuka et al. 2000a, Mikami, Otsuka and Yamato 2009, Panangadan and Talukder 2010), data trends (Qiao et al. 2012, You and Kim 2000), image pattern recognition (Turk and Pentland 1991, Ahonen, Hadid and Pietikainen 2006), neural networks (Lint, Hoogendoorn and Zuylen 2005b, Hinsbergen et al. 2011), etc. In this study, the average Euclidean distance between the current temporal-spatial traffic data and each data matrix with the same dimension from the historical data set is calculated using Equation (168) to represent a dissimilarity measure. Other advanced methods can be adopted to increase the matching speed and accuracy and are being considered as part of future research efforts.

$$d(c,h) = |M(c,L) - M(h,L)|/(L \times N). \tag{168}$$

where $M(c,L)$ and $M(h,L)$ represent the traffic data of the current and historical time intervals, respectively; and $d(c,h)$ is the average Euclidean distance between the traffic speed matrix data of different time intervals.

A small dissimilarity measure indicates that the matching historical data are similar to the current traffic pattern. Consequently, several candidates are selected according to the ascending order of the dissimilarity measure. Here, the maximum number of candidates is denoted by K, and the minimum acceptable dissimilarity is defined by d_{MIN}. The set of candidates H_c is selected as

$$\begin{aligned} H_c &= \{h_1, h_2, \cdots, h_{K'}\} \\ \text{where} \quad h_i &= \arg\min d(c,h) \\ d(c,h_i) &\leq d(c,h_{i+1}) \\ K' &= \max\{i \mid i \leq K, d(c,h_i) \leq d_{MIN}\} \\ |h_i - h_j| &\leq \varepsilon, \quad i \neq j \end{aligned} \tag{169}$$

where h_i is the selected candidate from the historical data set; K' denotes the resulting number of the selected candidates; and ε is used to avoid selecting adjacent candidates from the same day in the history data. The selected candidates represent the best matching to the current traffic status and will be used to calculate future travel times.

4.3 Dynamic Travel Time Prediction

The future dynamic travel times on the current day can be calculated based on the selected historical candidates. Considering the stochastic nature of a traffic system, the travel time prediction problem can be recognized as a time series prediction for nonlinear dynamic (chaotic) systems (Basharat and Shah 2009, Ikeguchi and Aihara 1995). The future traffic state for the current day can be predicted by the subsequent traffic state of each candidate from the historical data set. The linear combination of each candidate's subsequent traffic state is used to predict the future traffic status, and the corresponding weight is defined as the inverse of the dissimilarity measure of each candidate. The prediction traffic state starting from time interval $c+p$ is obtained as

$$M(c+p) = \sum_{i=1}^{K'} w(h_i) \cdot M(h_i + p) \tag{170}$$

$$w(h_i) = \frac{d(c,h_i)^{-1}}{\sum_{i=1}^{K'} d(c,h_i)^{-1}} \tag{171}$$

where $M(h_i+p)$ represents the p steps ahead subsequent traffic state for i^{th} candidate; and $w(h_i)$ denotes the weight of ith candidate data.

The next step is to calculate the dynamic travel time based on the subsequent traffic state of each candidate. Dynamic travel time is the actual, realized travel time that a vehicle could experience during a trip. If a vehicle leaves a trip origin at the current time, the roadway speed will not only change across space but also across time during the entire trip. Therefore, the traffic state evolution over space and time is considered in our approach, as shown in Figure 8 to compute dynamic travel times. The speed values of the shaded cells are used to compute dynamic travel times. In this paper, the traffic state is assumed to be homogenous within each cell. Therefore, the trajectory slope, which represents the traffic speed, is a constant value in each cell. Assume that the trip starts from time interval t_n. In this way, once the vehicle enters a new cell, the trajectory within this cell can be drawn as the straight dotted line in Figure 8, with the slope value equal to the traffic stream speed. Finally, the dynamic travel time can be calculated when the trajectory reaches the downstream boundary of the last freeway section (destination).

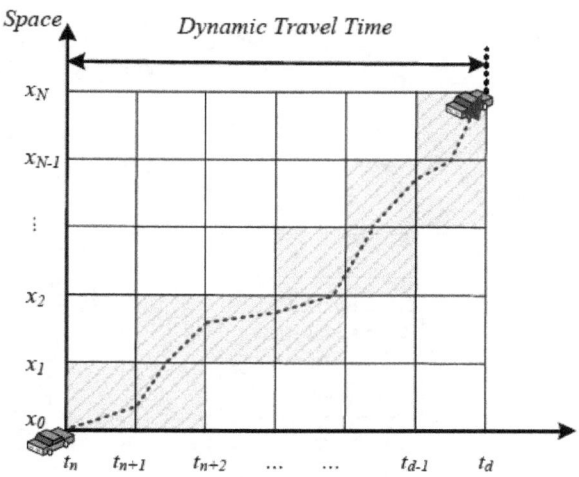

Figure 8. Illustration of dynamic travel time.

The procedure for computing dynamic travel times is shown in Figure 9. Consequently, the dynamic travel time of each subsequent candidate can be obtained and the corresponding weight (recurrent probability) is defined by the dissimilarity measure of Equation (172). Finally, the travel time distribution of the future trip can be represented as

$$TT(c+p) = \{TT(h_i+p), w(h_i) \mid i=1,\cdots,K'\} \tag{172}$$

where $TT(c+p)$ represents the dynamic travel time starting from time interval $c+p$; and $TT(h_i+p)$ denotes the subsequent travel time of ith candidate, according to the calculation as shown in Figure 9. The travel time prediction result can also be calculated as the average value using

$$\overline{TT}(c+p) = \sum_{i=1}^{K'} w(h_i) \cdot TT(h_i+p) \tag{173}$$

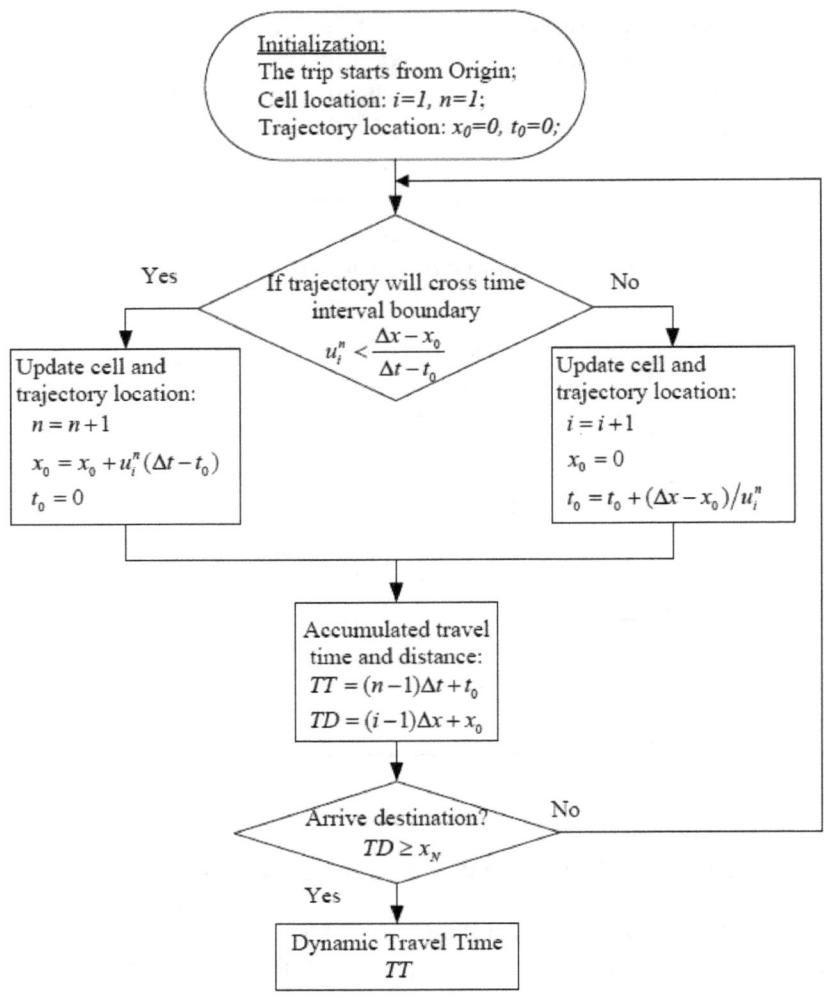

Figure 9. The flow chart of dynamic travel time calculation.

An illustration of how to use current traffic status to find historical candidates and then predict travel time is shown in Figure 10. The testing day is August 4, 2010 and the current time is 16:00 p.m. The current traffic status is represented by a matrix between the first and second vertical lines along all the sections. The template matching algorithm is implemented to find three historical data sets (May 28, 2010, August 2, 2010 and August 3, 2010) with a similar traffic pattern to the current traffic status. The image strips between the first and second lines on all four days include very similar congestion patterns (represented by red color) around the 50th section. Consequently, the future spatiotemporal traffic states from the selected candidates are used to represent the future traffic status of current day. This deduction can be supported by the fact that the image strips between the second and third lines on four days are very similar.

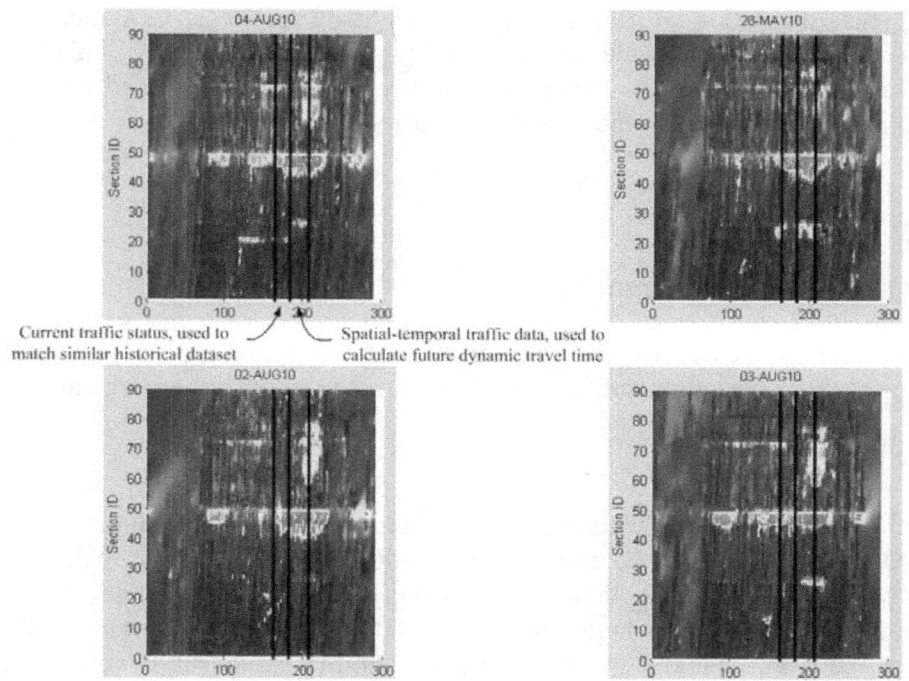

Figure 10. Samples of current traffic status and selected candidates.

4.4 Revised Algorithm

Based on the first case study of using the proposed approach to predict travel time, several improvements have been adopted to our algorithm to improve the prediction accuracy. The revised prediction algorithm will be tested on the second and third case studies, and the detailed information of the revised algorithm is presented below:

1) The value of L is correlated to the current instantaneous travel time instead of using a fixed value for every time interval. For instance, a small value of L is used during uncongested time intervals and a large value is used for congested time intervals. The criterion is computed based on the value of the instantaneous travel time.

2) The selected candidate number K' is also a variable correlated to the instantaneous travel time. During uncongested conditions, K' is set to be a large value in order to obtain a smoothed aggregation value from historical candidates. During time periods of high congestion, a smaller number of candidates is used since there are limited high congested data in our existing database. This may not be a problem in the future when the historical data set increases.

3) During the process of template matching for each day in the historical data set, the data slice of the best fit (least matching cost) is selected. Moreover, the searching range is constrained to a fixed 2-hour window around the current time c. Therefore, only one data slice is selected for each day in the historical data set and, finally, K' candidates are selected from all data slices.

4) During the calculation of dissimilarity, a weight parameter is introduced to utilize the length of each section to scale the corresponding dissimilarity value. This is based on the fact that the same level of congestion on a long section compared to a short section should have more effect on the total dissimilarity calculation.

5 Findings of Traffic Estimation

In this chapter a series of relevant numerical tests on traffic modeling, data fusion and travel time data representation are presented. But, first traffic datasets used in this research are introduced.

5.1 Traffic Datasets

In this research four standard traffic datasets prepared and made available under Next Generation SIMulation (NGSIM) project will be used (FHWA 2006). These datasets provide a rich source of accurate and very detailed traffic data over a variety of facilities located in California and Georgia. These datasets contain high quality (resolution equal to one-tenth of the second) observations of the type, position, speed, and acceleration of every single vehicle that has been part of the traffic stream in the segment under study. These datasets include the following:

5.1.1 The I-80 Data Set

A data set representing 45 minutes of data collected during the afternoon peak period on a segment of Interstate 80 in Emeryville (San Francisco), California. The data set consists of detailed vehicle trajectory data, wide-area detector data and supporting data needed for behavioral algorithm research on a merge section of eastbound I-80.

5.1.2 The US 101 Data Set

A 45-minute freeway data set representing traffic flows on a segment of U.S. Highway 101 (Hollywood Freeway) in the Universal City neighborhood of Los Angeles, California. The dataset represents vehicle trajectory data on a 2,100 foot, six lane segment of southbound U.S. 101,. The merge/weave section represented in the data includes an on-ramp off-ramp connected by an auxiliary lane. The dataset consists of detailed vehicle trajectory data, wide-area detector data and supporting data needed for behavioral algorithm research.

5.1.3 The Lankershim Data Set

A 30-minute arterial dataset representing traffic flows on a segment of Lankershim Boulevard in the Universal City neighborhood of Los Angeles, California. The dataset represents bi-directional vehicle trajectory data for an approximate 1,600-foot, three-to-four lane arterial segment of Lankershim Boulevard. The dataset consists of detailed vehicle trajectory data, wide-area detector data and supporting data needed for behavioral algorithm research.

5.1.4 The Peachtree Data Set

A 30-minute arterial dataset representing traffic flows on a segment of Peachtree Street in the Midtown neighborhood of Atlanta, Georgia. The dataset represents bi-directional vehicle trajectory data for an approximate 2,100-foot, two to three lane arterial segment of Peachtree Street, including one stop-controlled intersection and four signal-controlled intersections with permitted/protected left turns. The dataset consists of detailed vehicle trajectory data and supporting data needed for behavioral algorithm research.

In the rest of this chapter results of experiments on US-101 dataset are presented.

5.2 Traffic Modeling

In this study, both FDM and FEM solution methods presented in chapter 4 are applied to the US-101 dataset. For this purpose, the length of the highway is broken into ten cells, each 210 feet (~64 meter) long, and solution is updated in two second time intervals. Also, based on separate investigations, the parameters in Greenshields' model, that is free flow speed and jam density, for this dataset have been estimated at 66 mph and 108 vehicles per mile per lane, respectively. Thus, for this case, we obtain the following linear speed-density relationship.

$$v\ [mph] = 66 - 0.611 \times k\ [vpmpl], \quad (R^2 = 0.13) \tag{1}$$

Statistical examination of error measures presented in Table 2 reinforces the notion that both FDM and FEM have an excellent performance in solving the velocity based continuum traffic model presented here. The bias in FDM and FEM solutions varies between one and three miles per hour. However, the mean absolute speed errors are in the four to six miles per hour range. Based on the results in Table 2, it seems that FDM has slightly a better performance compared to FEM. The results, however, are too close to make it possible for a general conclusion. Further comparison of these solution methods under different cell/element configurations and with different datasets may be necessary before a conclusive result can be obtained.

Table 2. Overall performance of the approximate solution methods.

Time Interval	Solution Method	Error (mph)					Absolute Error (mph)				
		Mean	Std. Dev.	%-ile 25	%-ile 50	%-ile 75	Mean	Std. Dev.	%-ile 25	%-ile 50	%-ile 75
7:50am-8:05am	FDM	2.1	6.3	-2.7	2.5	6.8	5.5	3.7	2.6	5.1	7.9
	FEM	2.7	6.4	-1.5	3.0	7.5	5.7	4.0	2.4	5.1	8.4
8:05am-8:20am	FDM	1.3	5.2	-1.9	0.8	4.5	4.1	3.5	1.3	3.0	6.1
	FEM	1.3	5.9	-2.4	0.8	4.9	4.6	3.8	1.6	3.6	6.9
8:20am-8:35am	FDM	1.7	5.6	-1.9	1.2	5.0	4.4	3.8	1.5	3.4	6.1
	FEM	1.7	6.2	-2.6	1.4	5.7	5.0	4.0	2.0	4.1	7.0

Figure 11 depicts the variations in mean absolute speed errors over time as a result of using either solution method in the first 15 minute time interval. This shows that generally mean absolute errors are less than 10 miles per hour at any given time. The overall bias and mean absolute errors the solution methods are in the one to three and four to six miles per hour range, respectively, as demonstrated in Table 2.

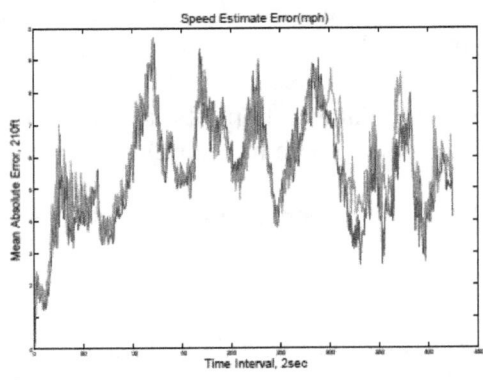

7:50a.m.-8:05a.m.

Figure 11. Mean absolute errors of the estimated speeds at elemental nodes of US-101 (blue and red lines represent errors from FDM and FEM methods, respectively)

Also, it was observed that both methods are capable of reproducing major shockwaves when they pass through both upstream and downstream of the segment. Smaller shockwaves initiating somewhere in between the two end boundaries have not been traced properly in the real-time approximate solutions. These shockwaves are reflecting higher order phenomena which obviously cannot be captured using a first-order model such as LWR model.

However, it seems that speed estimates obtained by solving LWR model may be further enhanced by incorporating additional traffic measurement sources in a stochastic state-space model framework. This may be conceived as a real-time estimation process which effectively combines LWR solutions with other direct or indirect speed measurements. Particularly, such an estimation process would be more beneficial in cases where measurement errors are significantly less than model errors. For instance, an additional loop detector in the middle of the segment or probe data could be used for this purpose. However, given the 0.4 mile length of this segment, it may not be realistic to assume in practice an additional loop detector will be in place for this purpose. Alternatively, assimilation methods to incorporate vehicle re-identification data into estimation process may prove as beneficial.

5.3 Traffic Data Fusion

The results given in traffic modeling section were based on solving a traffic dynamic model represented as a partial differential equation with accurate initial and boundary value conditions. However, in reality boundary speed measurements are not accurate. In this segment results of tests performed on the same problem with corrupted model and measurements are presented. A white noise error with standard deviation equal to five miles per hour is added to the model and boundary speed measurements.

In addition, the effect of probe data on estimation quality is incorporated into the model using random speed measurements from various points in the space-time solution domain. These internal speed measurements are corrupted by adding another white noise with three miles per hour standard deviation to the observed speeds.

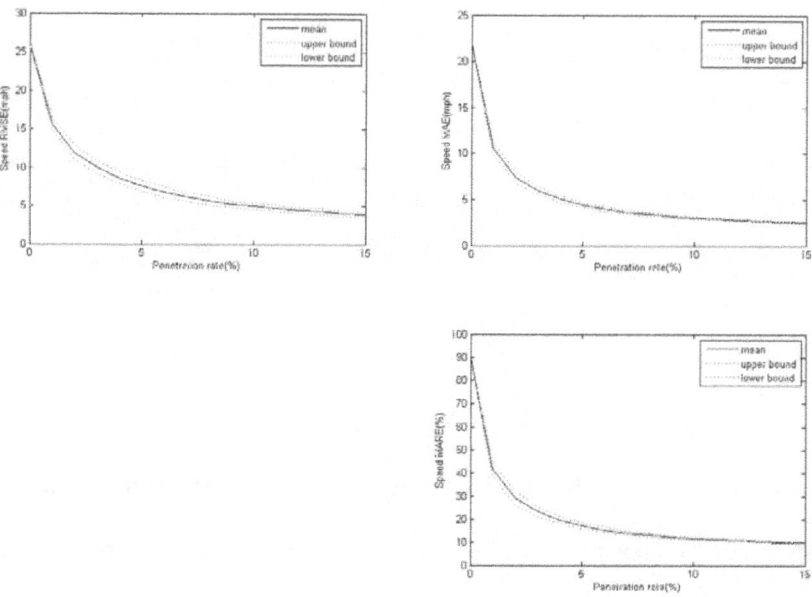

Figure 12. DKF speed estimate qualities at different penetration rates.

Figure 12 presents the error measures obtained by applying discrete Kalman filtering (DKF) to the corresponding state-space model at different levels of probe penetration inside the traffic stream. At each penetration rate 30 replications of problem are solved. A mean and 95 percent confidence band curve for each error measure are presented. These results also verify Work et al. (2008) observation that five percent penetration rate is a critical value below which estimated speed error will sharply decrease with any additional probe data, while at larger penetration rates errors tend to decrease much more slowly with increasing penetration rates. When only boundary speed data are used (0% penetration rate), RMSE of the estimates is about 25 miles per hour which is consistent with the added white noise to the measurements and the model. Average absolute and relative absolute errors (MAE and MARE) in speed estimates at penetration rates below five percent are observed to be larger than 5 miles per hour and 20%, respectively. Interestingly, at zero penetration rate (Eulerian data alone) the MAE is at 22 miles per hour and the MARE stands at 90%. This shows the great value of fusing internal data (Lagrangean) with existing boundary speeds (Eulerian) to improve the overall estimation quality.

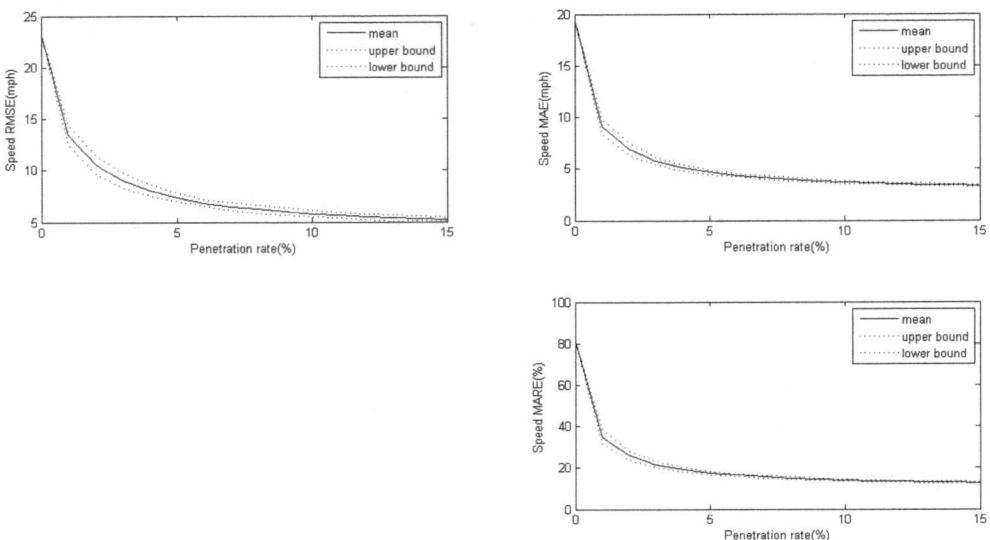

Figure 13. H_∞ speed estimate qualities at different penetration rates.

Figure 13 presents the speed error measures when H_∞ estimation method is applied. This figure corresponds to Figure 12.

5.4 Travel Time Model and Finite Difference Schemes

Spatiotemporal speed and travel time observations along with their boundary values are presented in Figure 14. The boundary values may be considered as accurate measurements obtained from a set of imaginary loop detectors accompanied with license plate matching cameras installed at the two boundaries of the segment.

Two scenarios are considered to evaluate the impact of discretization on solution quality of the proposed finite difference schemes. First, the space time domain is broken into 210 feet (~64 meter) long cells with two second updates. Second, a slightly coarser 420 feet (~128 meter) by 4 second discretization is adopted. It should be noted that in both cases the $\frac{\Delta x}{\Delta t}$ ratio is equal to 105 feet per second which is higher than free flow speed satisfying the stability condition.

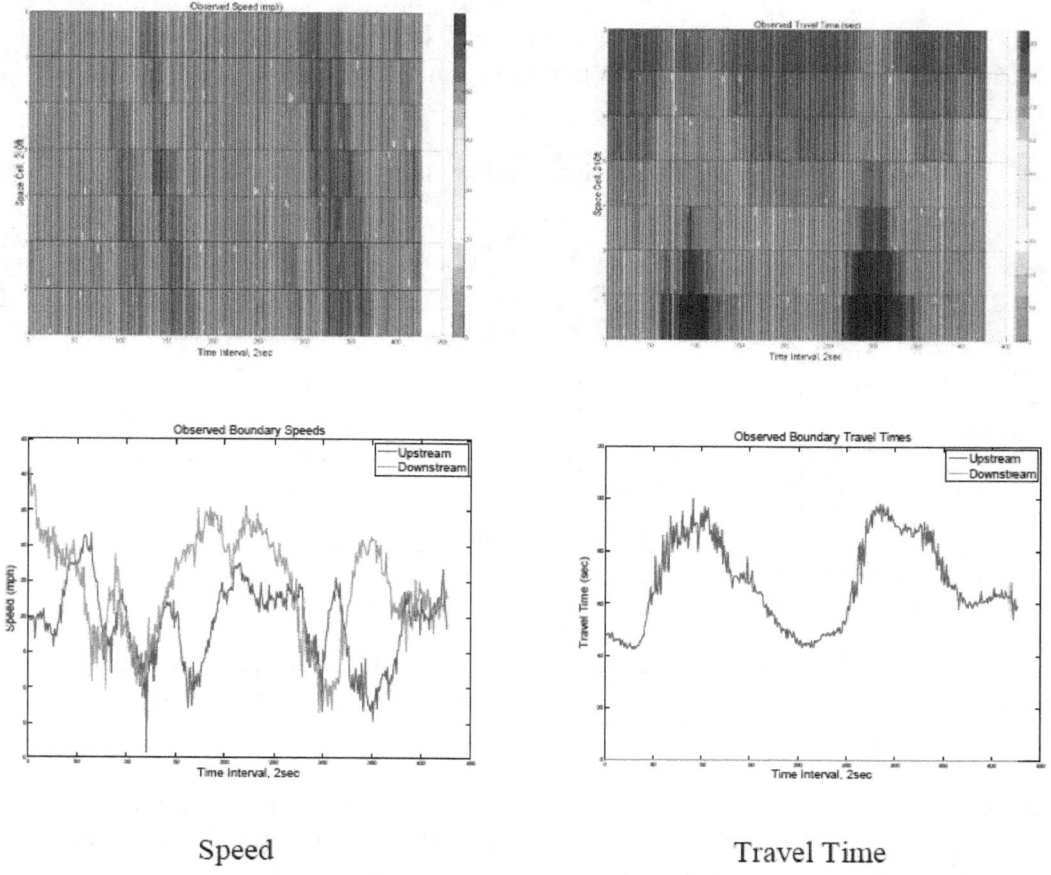

Figure 14. Spatiotemporal and boundary observations on US-101 mainlines. (upstream and downstream values are represented by blue and red lines, respectively)

Table 3 presents the quality measures of the speed estimates obtained from solving first order velocity based continuum traffic flow model (LWR-v) model with Greenshields' flux using Godunov finite difference scheme on a space-time plane (Sadabadi and Haghani, 2011).

Table 3. Speed estimation quality

Time Period	Resolution	Error (mph)					Absolute Error (mph)				
		Mean	Std. Dev.	%-ile			Mean	Std. Dev.	%-ile		
				25	50	75			25	50	75
0750am-0805am	210ft*2sec	2	6	-3	2	7	6	4	3	5	8
	420ft*4sec	3	5	-1	3	7	5	4	2	4	7
0805am-0820am	210ft*2sec	1	5	-2	1	4	4	4	1	3	6
	420ft*4sec	1	5	-2	0	4	4	3	1	3	6
0820am-0835am	210ft*2sec	2	6	-2	1	5	4	4	2	3	6
	420ft*4sec	2	5	-2	1	5	4	4	1	3	6

Solutions belonging to all three 15 minute periods in the morning rush hour are presented. At both resolutions the estimates are comparable in terms of their accuracy. The bias in speed estimates varies between one and three miles per hour, while the mean absolute speed errors are in the four to six miles per hour range.

Table 4. Travel time estimation quality using observed speeds

Time Period	Resolution	Error (sec)					Absolute Error (sec)				
		Mean	Std. Dev.	%-ile 25	50	75	Mean	Std. Dev.	%-ile 25	50	75
		FTBS Scheme									
0750am-0805am	210ft*2sec	1	2	0	1	3	2	2	1	2	3
	420ft*4sec	2	2	1	2	3	2	1	1	2	3
0805am-0820am	210ft*2sec	1	3	0	0	1	2	3	0	1	2
	420ft*4sec	1	2	0	1	2	2	2	0	1	2
0820am-0835am	210ft*2sec	1	3	-1	0	2	2	2	0	1	3
	420ft*4sec	1	2	0	1	2	2	2	1	1	3
		Godunov Scheme									
0750am-0805am	210ft*2sec	4	5	1	3	6	5	4	1	3	7
	420ft*4sec	4	3	1	3	5	4	3	2	3	5
0805am-0820am	210ft*2sec	2	6	0	1	4	4	5	1	2	6
	420ft*4sec	1	4	0	1	3	3	3	1	2	4
0820am-0835am	210ft*2sec	2	6	-1	1	5	4	5	1	2	6
	420ft*4sec	2	4	0	1	4	3	3	1	2	4

Even though the speed accuracy seems to be high enough for travel time estimation in this case, it is desirable to have a measure of the impact of speed accuracy on travel times estimates. To this end, travel times were estimated based on both observed and estimated speeds. At the same time, this will provide a basis for evaluating the accuracy of proposed travel time estimation schemes independent of the quality of speeds used in such estimations.

Quality measures of travel time estimates obtained based on observed speeds using FTBS and Godunov schemes are presented in Table 4. Travel time estimates obtained applying FTBS scheme have a one to two second bias while their mean absolute error is consistently about two seconds in all three time periods and under both discretization scenarios. However, the Godunov scheme has resulted in slightly higher bias varying from one to four seconds and higher mean absolute errors between three to five seconds under these circumstances. Comparing the first and third quartiles with the median of error distribution in both methods suggests that errors are generally symmetrically distributed. Despite the fact that Godunov scheme has generated larger errors in this case it seems that as congestion increases its performance improves as opposed to FTBS scheme which has displayed a rather uniform performance throughout this case.

Figure 15 illustrates the spatiotemporal travel time estimates at 210 feet by 2 second resolution using observed speeds and their mean absolute error variation over time period 08:05am to 08:20am in the morning rush hour.

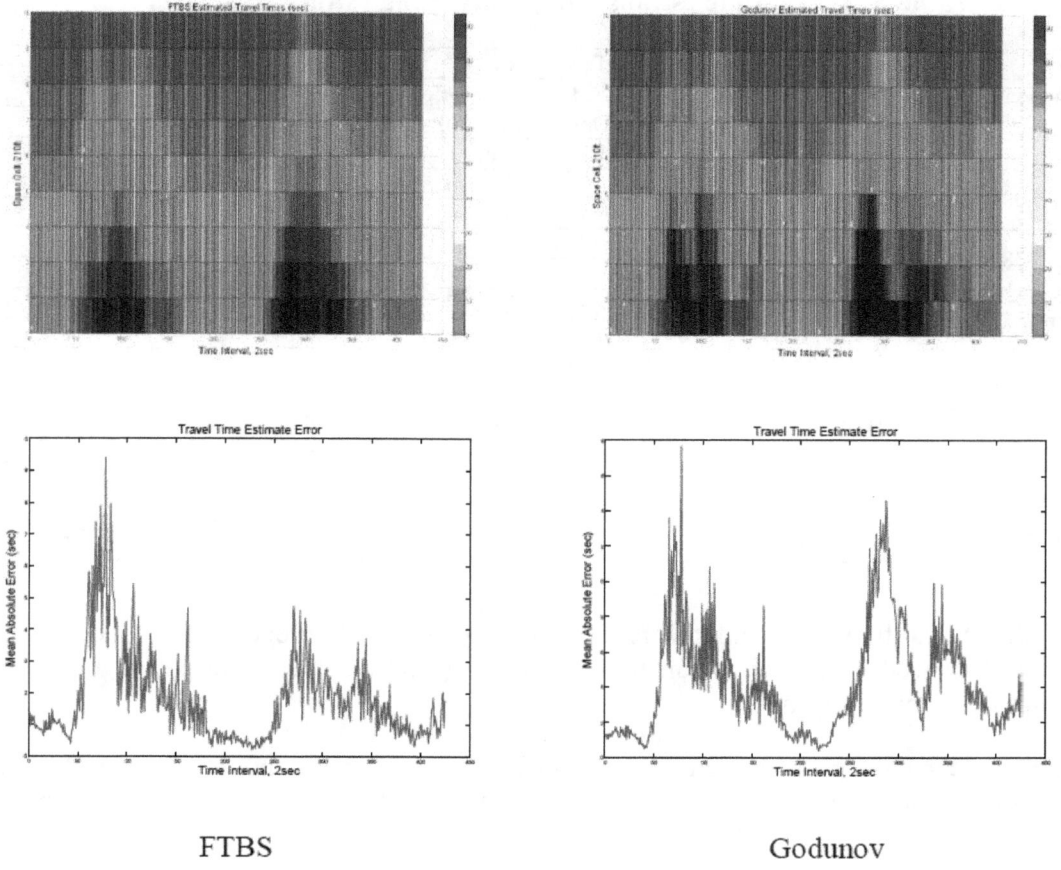

Figure 15. Travel time estimates based on observed speeds on US-101 mainlines (top: spatiotemporal travel time estimates, bottom: mean absolute error)

However, it is most interesting to see how the two schemes have performed when speed estimates rather than observations are used in the estimation process. Table 5 presents quality measures of travel time estimates based on estimated speeds using FTBS and Godunov schemes. Table 5 shows that travel time estimates obtained applying FTBS scheme have a six to nine second bias, while their mean absolute error is seven to ten seconds in all three time periods and under both discretization scenarios. Table 5 also shows that, generally speaking, Godunov scheme has resulted in similar bias varying from six to eight seconds and comparable mean absolute errors between six to nine seconds under these circumstances. Again judging by the relative extent of the middle quartiles of the error distributions it seems that error distributions are symmetric.

Table 5. Travel time estimation quality using estimated speeds

Time Period	Resolution	Error (sec)					Absolute Error (sec)				
		Mean	Std. Dev.	%-ile 25	%-ile 50	%-ile 75	Mean	Std. Dev.	%-ile 25	%-ile 50	%-ile 75
FTBS Scheme											
0750am-0805am	210ft*2sec	7	5	4	6	10	7	4	4	7	10
	420ft*4sec	7	4	4	6	9	7	4	4	7	9
0805am-0820am	210ft*2sec	7	7	2	5	12	8	6	2	6	12
	420ft*4sec	6	6	2	5	10	7	6	2	5	10
0820am-0835am	210ft*2sec	9	9	2	8	13	10	8	4	8	13
	420ft*4sec	8	8	3	8	12	9	7	4	8	12
Godunov Scheme											
0750am-0805am	210ft*2sec	7	5	4	6	10	7	4	4	7	10
	420ft*4sec	6	4	3	6	9	6	4	4	6	9
0805am-0820am	210ft*2sec	7	7	2	5	11	8	6	2	6	12
	420ft*4sec	6	6	1	4	9	6	5	1	5	9
0820am-0835am	210ft*2sec	8	9	2	7	12	9	8	4	8	12
	420ft*4sec	8	8	3	7	11	8	7	3	7	11

Figure 16 illustrates the spatiotemporal travel time estimates at 210 feet by 2 second resolution using estimated speeds and their mean absolute error variation over time period 08:05am to 08:20am in the morning rush hour.

In essence, three factors capable of affecting performance of the proposed schemes for travel time estimation that is speed accuracy, traffic congestion level and discretization level are considered in our tests.

First, accuracy of speeds used in such estimation has a direct effect on the quality of travel time estimates. Therefore, to provide a benchmark for comparison, two sets of travel time estimates based on observed and estimated speeds are reported in the numerical experiments. These experiments showed that proposed schemes are indeed very accurate when accurate speeds are used. In fact, under this scenario travel time estimates with MAPE of about 6% and 10% were obtained using FTBS and Godunov schemes, respectively. However, travel time as an integral of traffic pace suffers from the presence of error in underlying speeds. Even though the MAE of speed estimates was in four to six mile per hour range, travel time estimates based on them were disproportionately less accurate. Under these circumstances FTBS and Godunov schemes have almost similar performance in terms of the accuracy.

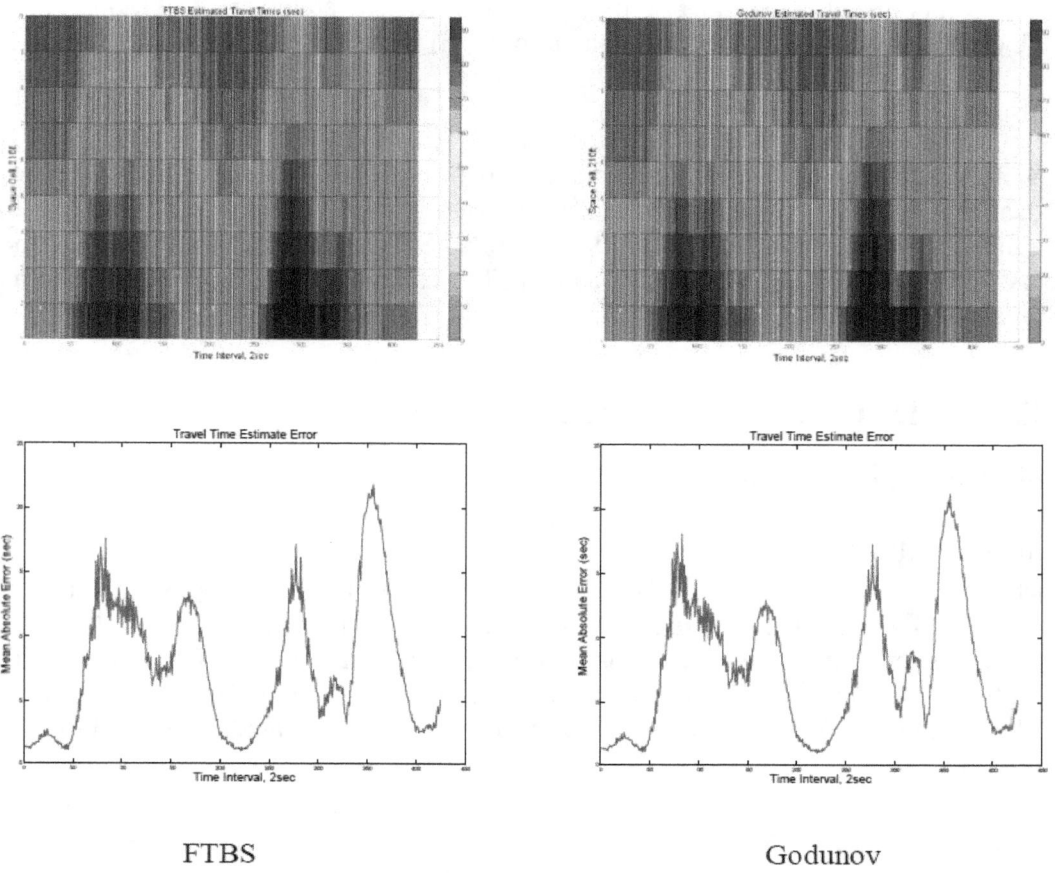

FTBS   Godunov

Figure 16. Travel time estimates based on estimated speeds on US-101 mainlines (top: spatiotemporal travel time estimates, bottom: mean absolute error)

Second, in our experiments congestion level and therefore the extent to which smoothness assumption is violated is varied in each of the three US-101 fifteen minute datasets. These datasets represent increasing congestion levels over time during a typical morning rush hour. Arguably, Godunov scheme exhibited a better performance at higher congestion levels as opposed to the FTBS scheme which quickly loses its edge as traffic conditions become more volatile.

Finally, specifics of the solution domain discretization which effectively determines the solution resolution is considered as an important factor which impacts the accuracy of proposed finite difference schemes. In the experiments two discretization stencils are examined. However, it should be noted that due to the limited length of the US-101 segment covered in the datasets as well as rather high free flow speed on this segment testing larger stencils was impossible. Despite these limitations, comparing errors at tested resolutions suggest that errors will decrease as the cell size increases. This observation is compatible with the fact that with increasing cell sizes variations inside the cells are smoothed out while discrepancies between adjacent cells become more pronounced. Therefore, it may be expected that at lower resolutions Godunov scheme will exhibit better performance compared to the FTBS scheme.

5.5 Summary

In this chapter, multiple high resolution data sets made available under NGSIM project are introduced as possible cases for numerical experimentation in this proposed research. Later, different numerical experiments on US 101 dataset of NGSIM project are reported. The reported experiments include applications in traffic data fusion, estimation, and travel time estimation using a proposed PDE model.

6 Findings of Dynamic Travel Time Prediction

6.1 Data Collection and Analysis

6.1.1 Acquire INRIX Data

Data reduction for the available INRIX data is an important task to be accomplished during this study. The acquired data set will be used to construct the travel database, and to develop and test the travel time prediction algorithm. After signing the Date Use Agreement (DUA) with RITIS and sending the request for INRIX data, two packages of probe data on I-64 and I-264 from the time period of October 2008 to November 2012 are obtained, as illustrated in Figure 17 (a) and (b), respectively.

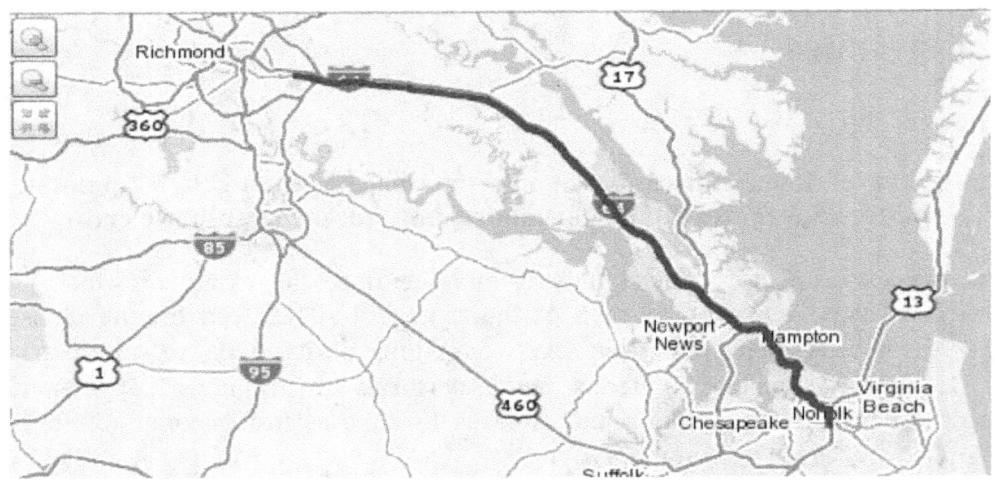

(a) INRIX data for I-64

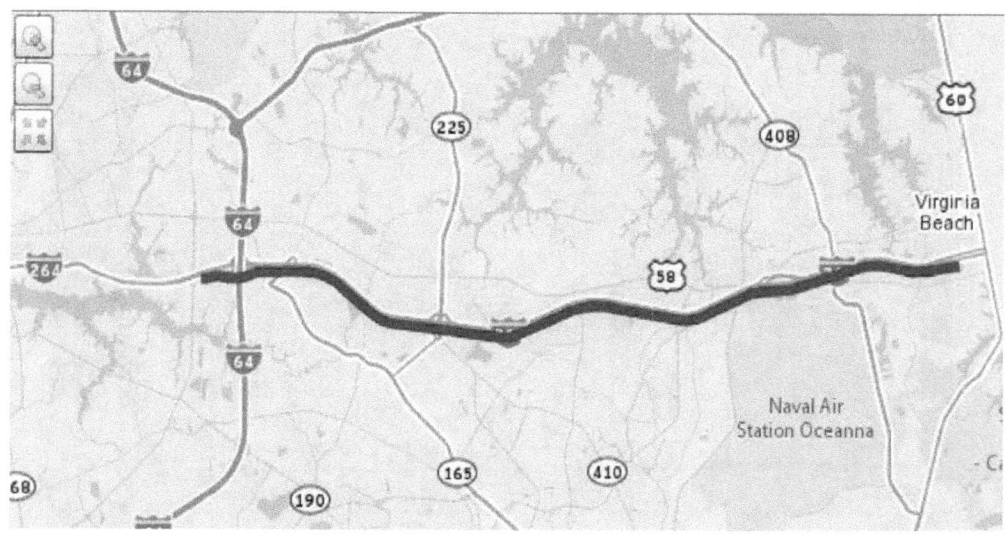

(b) INRIX Data for I-264

Figure 17. INRIX data for I-64 and I-264.

According to Figure 18, the aggregated speed data on the RITIS website demonstrate the data coverage information for I-64. The gray color indicates that the corresponding area has no data at all, and the green/yellow/red color indicates that traffic measurement is available. During the time period before October 7 2011, only the first 11 freeway sections include traffic data measurement. Those 11 sections are located east of I-295, which is out of the range of the study area. Therefore, the INRIX data for those sections cannot be used for this study. The speed data from other sections are available starting from 12:00 p.m. on October 6, 2011. However, the Hampton Roads Bridge-Tunnel (HRBT) still had no measurement until the end of January 2012. Finally, the full coverage data can be observed from the end of January 2012 through November 2012.

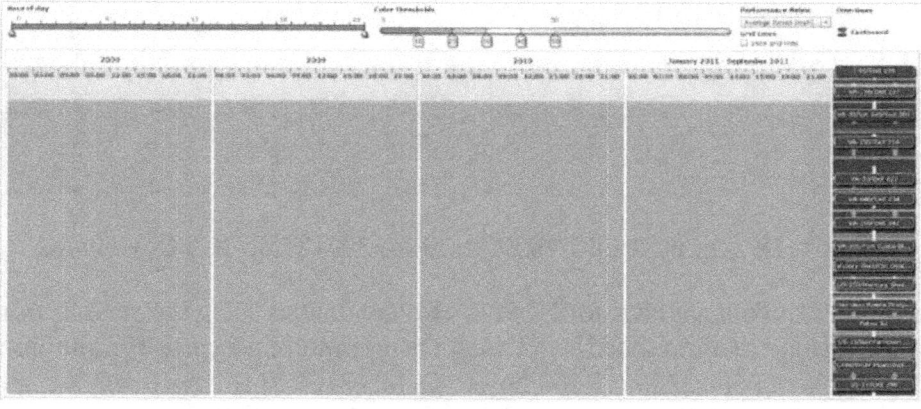

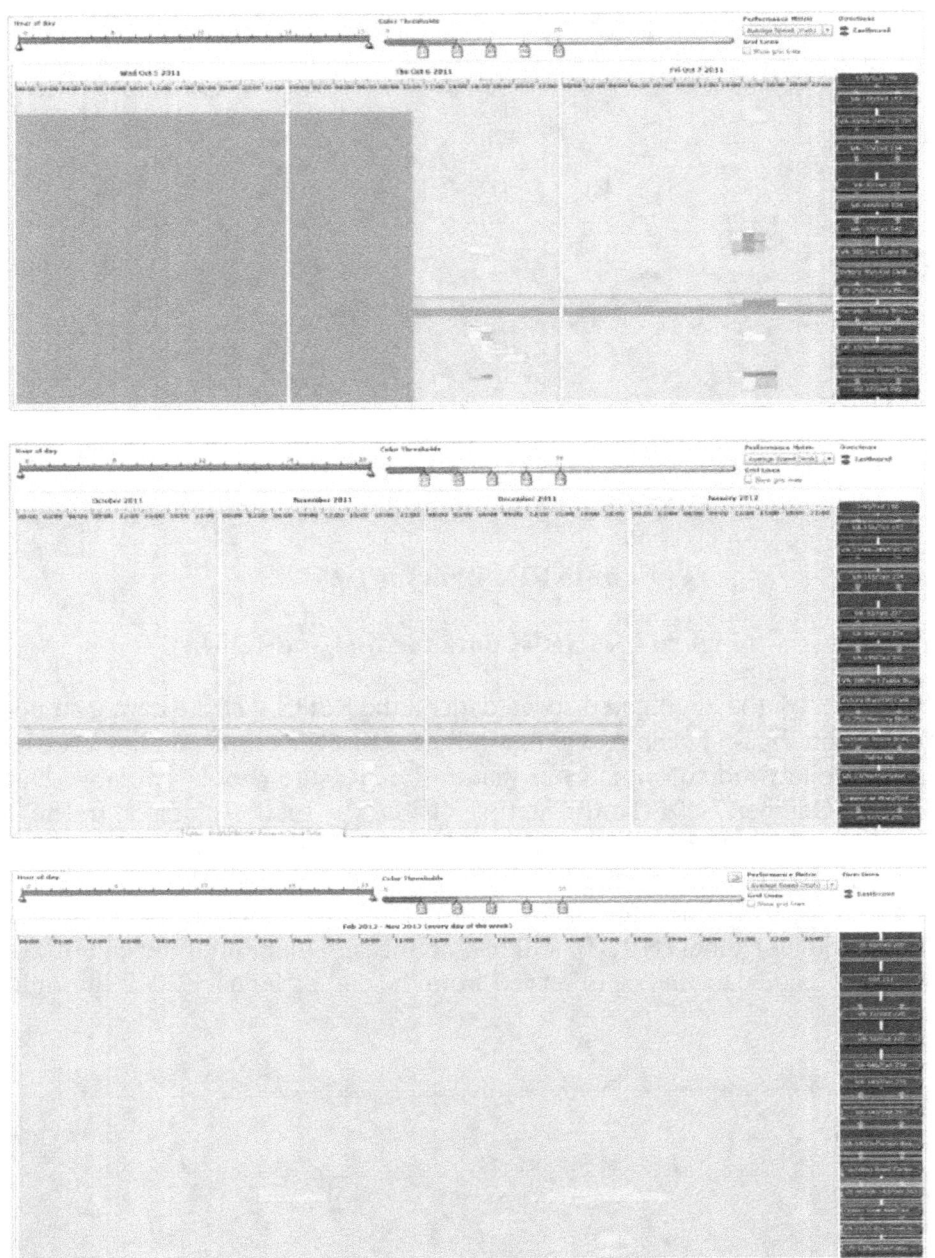

Figure 18. Aggregated INRIX data for I-64 from RITIS website.

The same procedure of data analysis for I-264 is also conducted using the service from the RITIS website, and the display of aggregated INRIX data is presented in Figure 19. Similar to I-64, no measurement data were collected on I-264 before October 6, 2011. Afterward, the speed data measurements have almost full coverage on I-264 from October 6, 2011 to November 2012.

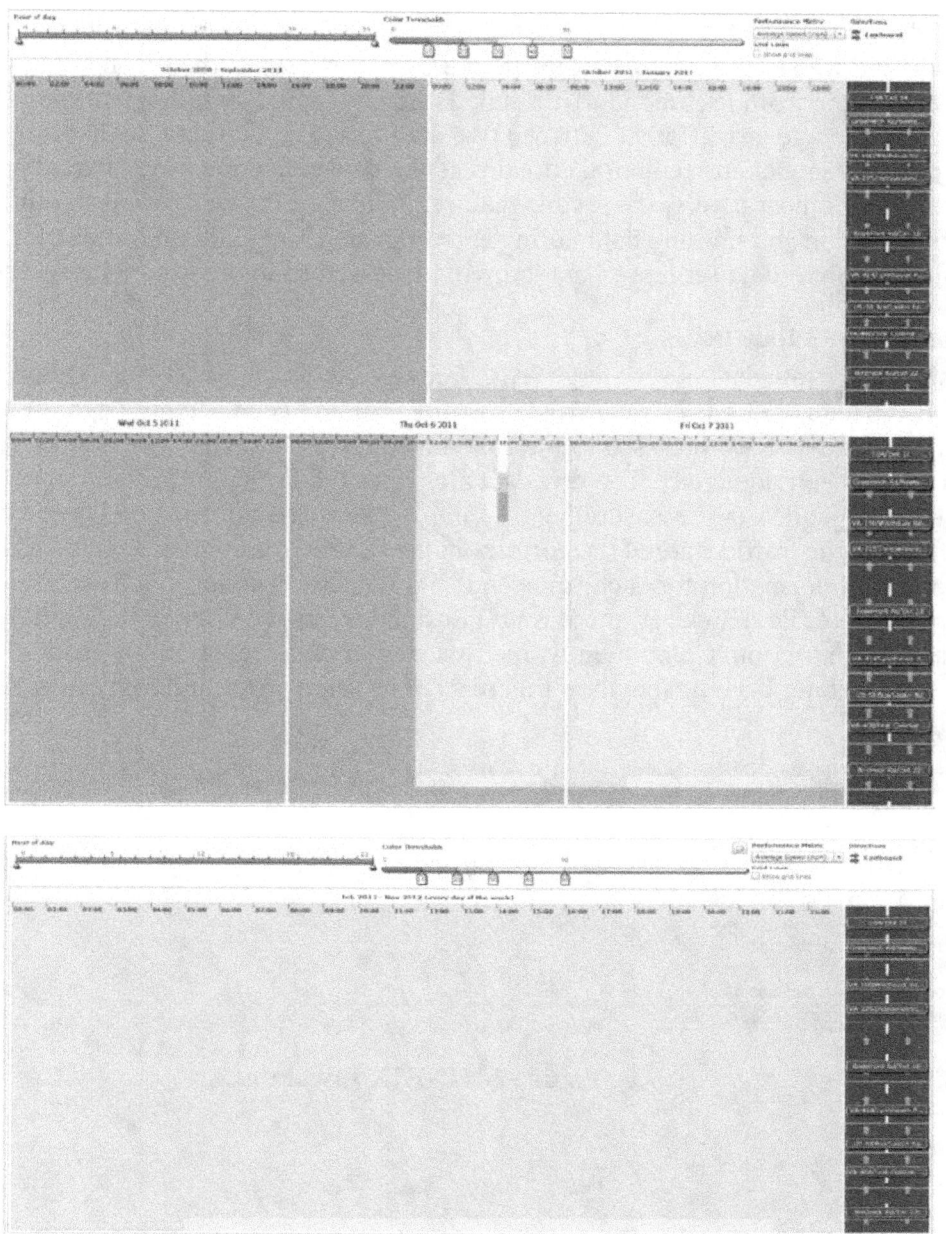

Figure 19. Aggregated INRIX data for I-264 from RITIS website.

As illustrated in Figure 18 and Figure 19, the ideal coverage of traffic measurements along I-64 and I-264 from Richmond to Virginia Beach starts on October 7, 2011, and ends on November 30, 2012. Considering that serious congestion mainly occurs along I-64 during that time, the summer holiday season (June, July, and August) is the main focus of this study. The summer 2012 data are not available when the proposed algorithm is tested; however, a third data package collected by INRIX and covering January to December of 2010 was added to this study. Therefore, the summer 2010 data can be used in this study to validate the algorithms of travel time prediction.

The INRIX data from October 7, 2011, to November 30, 2012 are used to construct the first data set along I-64 and I-264. The 2010 INRIX data are used to construct the second data set with the same freeway stretch from Richmond to Virginia Beach. It should be noted that the data resolution and coverage are different between two data sets. The first data set has a fine resolution over the spatial and temporal domains as the data collection time interval is usually around 1 minute. Comparatively, the second data set includes a coarse coverage with 5 minutes of data representation and missing data during early morning/late night and weekends. The detailed information of data set representation will be described in the following sections.

6.1.2 Problems with Raw Data

Since the size of the raw data is very large (e.g., freeway sections on I-64 from October 2011 to January 2012 include 3.33 GB raw data), Microsoft Excel or Access cannot open the data directly. SAS and MATLAB are used to filter the raw data to obtain the spatiotemporal average speed data of each individual day. The raw data are presented in Figure 20. Each row is generated by "tmc_code" and "measurement_tstamp." The speed information is used during this study to represent the traffic state of the corresponding roadway section at each time interval. The geographical information for each "tmc_code," which corresponds to a freeway section, is defined in a separate file. However, there is no information given to define the spatial relationship of each section. Consequently, the first step of data reduction is to sort all sections from west to east along I-64, as shown in Figure 21. The same procedure for sorting sections is adopted for I-264.

tmc_code 1	measurement_tstamp 2	speed 3	average_speed 4	reference_speed 5	travel_time_minu... 6	confidence_score 7	cvalue 8
110+04706	2011-10-06 11:58:46.0	58	60	63	1.821	30	100
110P04702	2011-10-06 11:58:46.0	59	59	62	0.707	30	100
110+04703	2011-10-06 11:58:46.0	59	60	63	0.326	30	100
110+04710	2011-10-06 11:58:46.0	60	59	62	0.976	30	100
110+04709	2011-10-06 11:58:46.0	65	61	63	0.689	30	100
110P04703	2011-10-06 11:58:46.0	60	61	64	0.432	30	100
110P04705	2011-10-06 11:58:46.0	62	60	63	0.532	30	100
110+04704	2011-10-06 11:58:46.0	62	59	64	1.009	30	100
110P04708	2011-10-06 11:58:46.0	66	61	64	0.345	30	99

Figure 20. INRIX raw data.

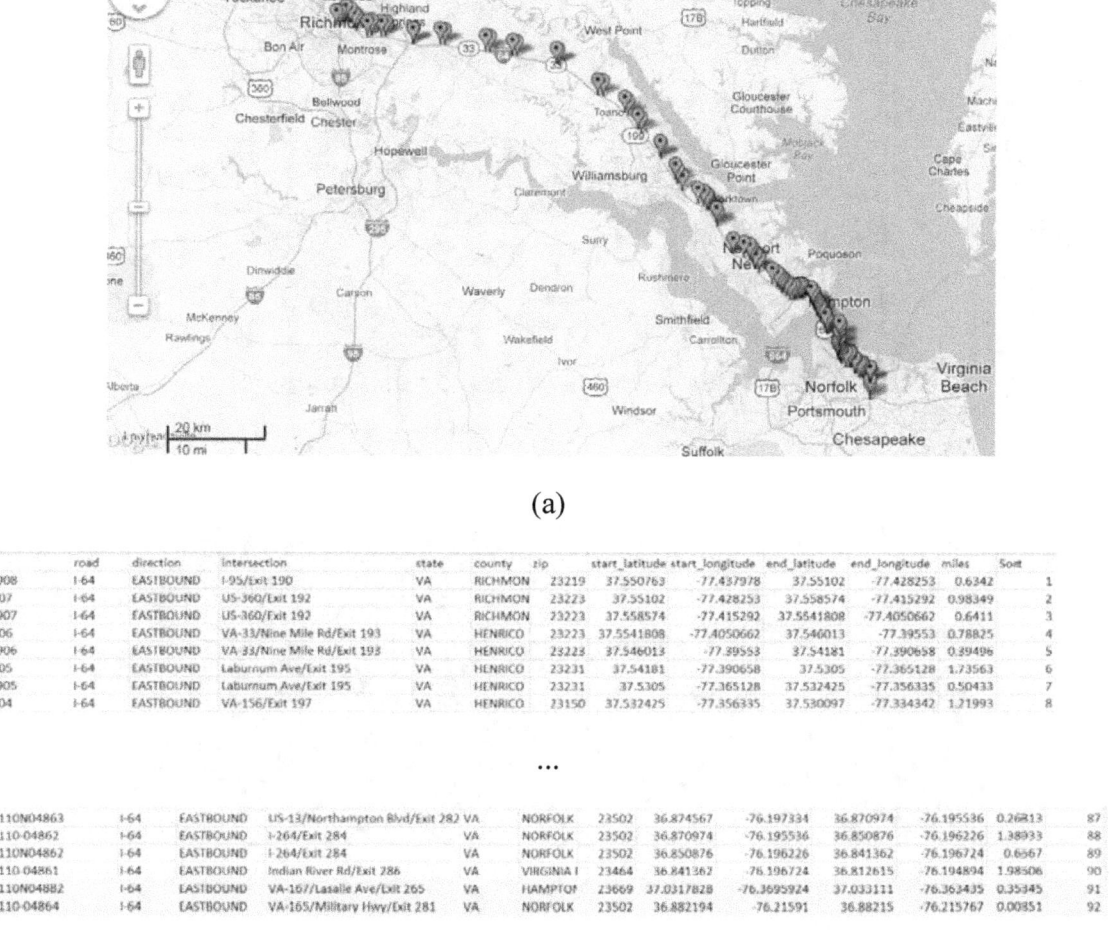

Figure 21. Sorted freeway sections along I-64.

A total of 92 sections are included in the mentioned I-64 data. However, two sections are not geographically consistent, as shown in Figure 22. The 91st section is overlaid by the 52nd section, and the 92nd section is overlaid by the 84th section. However, the 85th section shares a boundary with the 92nd section, not with the 84th section. Two overlaid sections are highlighted in yellow in Figure 21 (b). Both cases occur on the freeway ramps, and the research team believes that the traffic data for overlaid sections are collected to calculate the travel time leaving or coming from ramps on I-64. These cases will be used to adjust travel time computation accordingly to meet future needs.

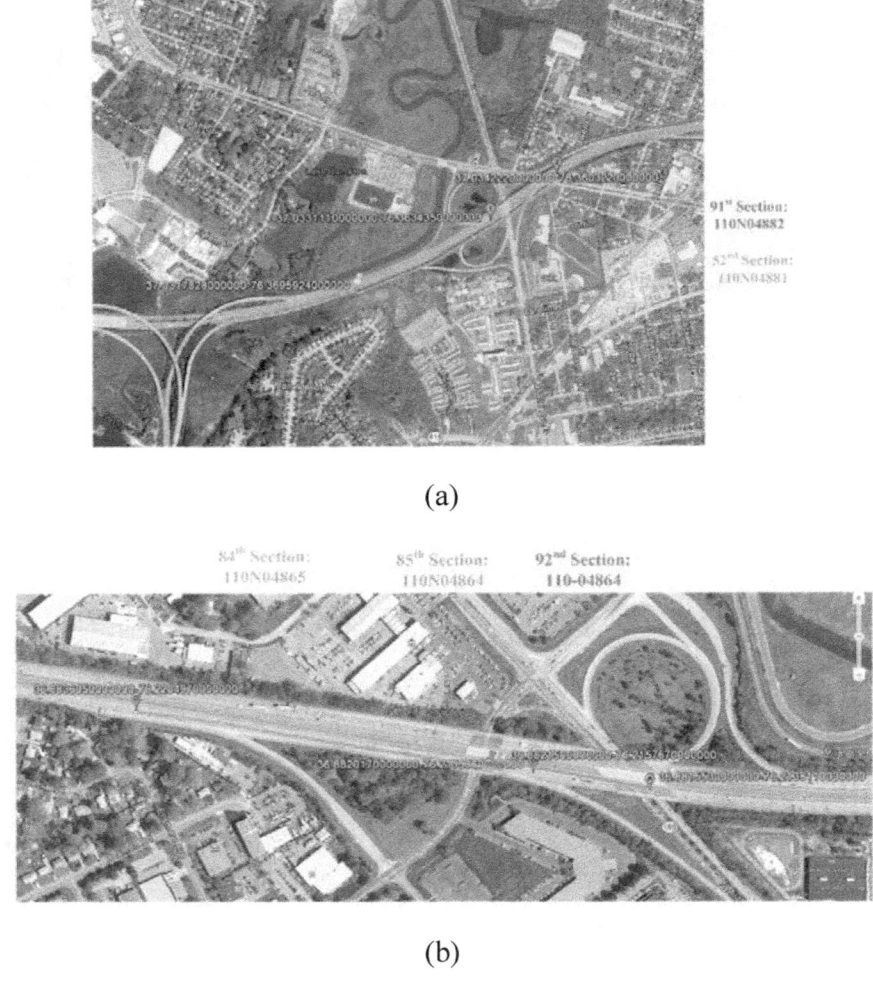

(a)

(b)

Figure 22. Geographically inconsistent sample sections.

The raw data set is loaded as Coordinated Universal Time (UTC); therefore, an adjustment is needed to change the time zone to Eastern Standard Time (EST). Specifically, the UTC from January 1 to March 13 and November 7 to December 31 during 2010 entails subtracting 5 hours, while the remainder of the 2010 data entails subtracting 4 hours. The raw data of 2011 and 2012 are adjusted accordingly. The probe data collected for irregular time intervals present another problem for data analysis. As illustrated in Figure 23, the measurement time interval varies between 59 seconds to 6 minutes. The irregular time interval will increase difficulty of data aggregation and data estimation during this study. Since most raw data collected after October 2011 are measured in a 1-minute interval, this part of the data is aggregated by calculating the average speed every 1 minute. The 2010 raw data are aggregated every 5 minutes.

```
12:01:00
12:02:00
12:06:00
12:07:00
12:09:00
12:10:01
12:14:00
12:16:00
12:18:00
12:21:00
12:22:00
12:28:00
12:30:00
12:33:00
12:36:00
12:39:00
12:46:00
```

Figure 23. Sample irregular time interval of raw data.

When INRIX created freeway segments to collect traffic data, the fact that I-64 and US-60 run concurrently for some distance was not considered. As a result, the collected traffic data for I-64 tunnel segments were listed under US-60 until January 20, 2012. Therefore, the research team requested (through RITIS) all past INRIX data for the US-60 highway in order to obtain the tunnel traffic data. The tunnel section was marked by "TMC 110+14251" in the US-60 data set with a section length of 3.716 miles. The available data we can obtain about this section are also collected between October 7, 2011 and January 20, 2012. This data set will be used to fill the gap of the tunnel segment on I-64 for the first data set. It should be pointed out that the tunnel segment on I-64 was separated by five sections after January 20, 2012, instead of just one section. Therefore, the collected traffic data for the tunnel area provide more detailed information after January 20, 2012.

Unlike the above information – where the I-64 tunnel segment is represented by one section between October 7, 2011 to January 20, 2012, and five sections afterward – the tunnel in the 2010 INRIX data set includes three sections corresponding to TMC identification numbers "110+14251," "110P14251," and "110+14252." The fact of different spatial segment compositions and various data collection time intervals will be considered accordingly to construct the travel database and to develop a prediction algorithm.

6.2 Travel Database Construction

After solving the aforementioned problems of raw INRIX data, a spatiotemporal traffic speed map can be generated for each individual day in order to construct the travel database. To serve the purpose of travel time prediction, the travel database should include full daily coverage of spatiotemporal speed information. Such information will be used in the prediction algorithm to obtain similar traffic patterns between the current day and historical data and to forecast future traffic information. After constructing the daily spatiotemporal speed map on the INRIX data set, the problem of missing data is discovered to be a serious problem. The missing data problem varies with different data sets. The existence of missing data and the corresponding solutions are presented in this section. A selected 37-mile freeway stretch from Newport News to Virginia

Beach is used to demonstrate the missing data problem. The same freeway stretch is used on the first case study to evaluate the proposed algorithm.

Considering the different data resolutions of spatial section composition and various data collection time intervals, two travel data sets are constructed as presented below.

- Travel Database 1: Based on INRIX data from October 7, 2011, to November 30, 2012; represented by 96 sections across the space axis and a 1-minute time interval across the time axis.

- Travel Database 2: Based on 2010 INRIX data; represented by 90 sections across the space axis and a 5-minute time interval across the time axis.

6.2.1 Generate Daily Spatiotemporal Speed Map for Travel Database 1

The freeway stretch used during this study includes sections along I-64 and I-264 from I-295 (east of Richmond) to I-264 (Virginia Beach). The entire 95-mile freeway is divided into 96 sections as shown in Figure 24. The section location is represented by mile posts starting from the first section in Richmond.

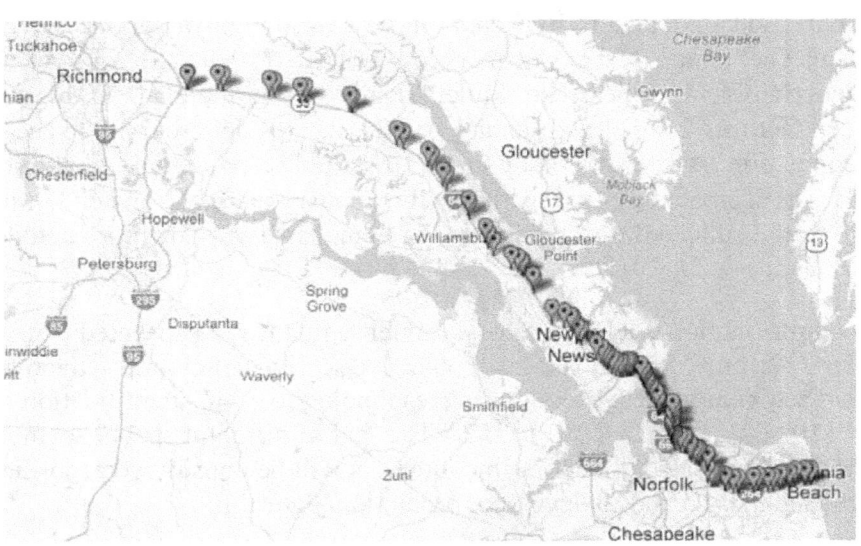

Figure 24. Freeway stretch for Travel Database 1.

Combining the data for I-64, I-264, and US-60, the speed data set along the entire freeway stretch is available from October 7, 2011, to November 30, 2012. After aggregating the speed data across space (section length) and time (1 minute), the daily temporal-spatial speed map is obtained. The speed map samples of typical weekday and weekend travel for Travel Database 1 are presented in Figure 25. The blue color represents free-flow speed, and the red color represents traffic congestion. It should be mentioned that the tunnel segment was covered by only one measurement for each time interval before January 20, 2012. This segment of data was filled by one section from US-60 as described in the Acquire INRIX Data section. For this case, the typical weekdays are represented as Figure 25 (a), (b), and (c). The typical weekend is represented as Figure 25 (d). Here we observe that the problem of missing data across the time space (between 1 to 20 minutes) occurs occasionally. The white strip areas of Figure 25 (a) and (c) demonstrate the missing data. The same approach of using neighboring data to estimate the missing measurement area is used for Travel Databases 1 and 2.

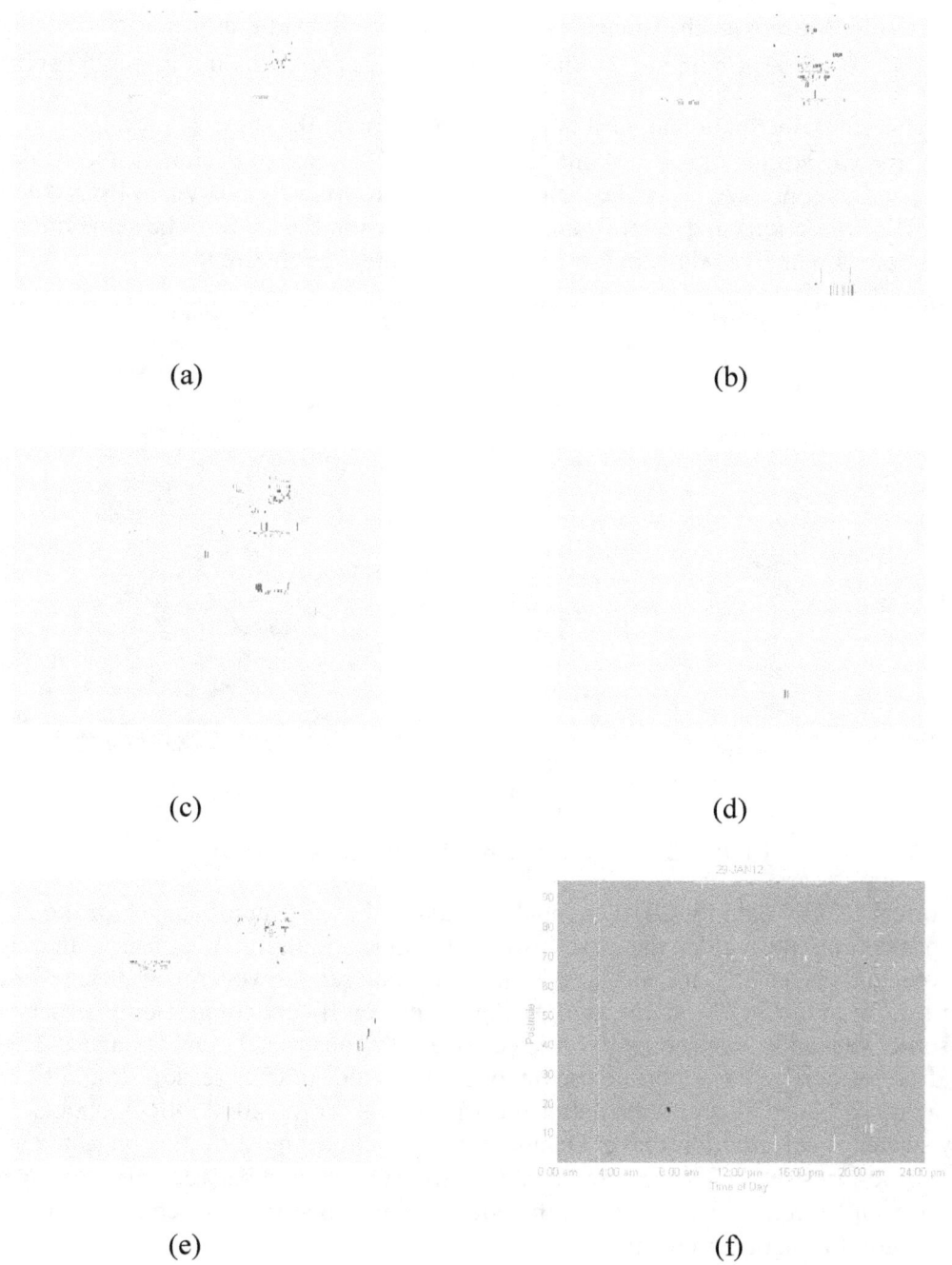

(a) (b)

(c) (d)

(e) (f)

Figure 25. Samples of daily spatiotemporal speed map for Travel Database 1.

Since the problem of missing data is more serious in Travel Database 2, detailed information of the estimation performance will be presented in the next section. Conversely, the tunnel segment was covered by five sections (five measurements for each time interval) since January 20, 2012. In this case, the typical weekday and weekend traffic maps are represented by Figure 25 (e) and (f), respectively. Moreover, the tunnel area includes more dynamic information compared with the blurred areas corresponding to the tunnel traffic status presented in Figure 25 (a) through (d).

6.2.2 Generate Daily Spatiotemporal Speed Map for Travel Database 2

The same freeway stretch along I-64 and I-264 from I-295 (east of Richmond) to I-264 (Virginia Beach) is used to construct Travel Database 2. The entire freeway is divided by 90 sections instead of the 96 sections in Travel Database 1, as shown in Figure 26. The same mile post representation as with Travel Database 1 is used to maintain consistency.

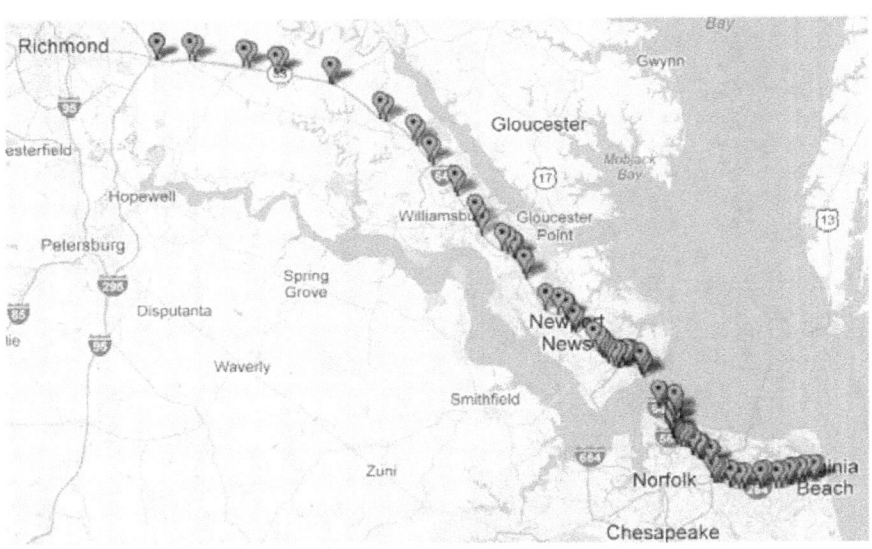

Figure 26. Freeway stretch for Travel Database 2.

The historical INRIX data for 2010 are analyzed using the same approach as for the 2011 data set except the time interval for the 2010 historical data is 5 minutes. We observe that traffic during weekends is usually uncongested in the 2011 data set. However, this differs for the summer season in the 2010 INRIX data, during which significant congestion is observed to occur during some weekends. Consequently, it is necessary to construct Travel Database 2 using 2010 INRIX data, especially if travel time predictions during the summer season should be investigated. To better illustrate the detailed traffic status for the 2010 INRIX data, a 37-mile freeway stretch is selected for Travel Database 2 that includes most of the congested areas along the entire freeway stretch. The selected freeway stretch is located between Newport News and Virginia Beach from the 33rd section to the 90th section. The same stretch is also used in the first case study for algorithm testing.

The data samples for typical weekday and weekend traffic occurring during June 2010 are presented in Figure 27. The figure illustrates a significant amount of missing data, especially for June 5 and June 6, 2010 (Saturday and Sunday). It appears from inspection of the data that the weekends involve more missing data than is the case for the weekdays, which may pose a problem, especially when making travel time predictions for weekend days. According to the speed map of Figure 27 (a) and (c), most missing data (white area) for a typical weekday occur between 21:00 p.m. and 5:00 a.m. (i.e., during the night and early morning hours). Normally there is low traffic volume during this time period, and free-flow speed could be assumed. However, sometimes the missing data also occur around a congested area (e.g., the speed map of Figure 27 e and g). Consequently, free-flow speed cannot be simply assumed for all missing data.

As demonstrated in the Literature Review section, various traffic data estimation algorithms are developed depending on the data resource. Since ramp traffic data are not available, greater errors will be introduced to macroscopic traffic models for estimating missing data. Alternatively, a statistical approach is employed here that uses temporal and spatial speed values around missing data. The average value of eight neighboring cells is used to estimate the missing speed data. Advanced approaches such as using kernel regression over spatial and temporal coordinates can be considered in the future. The samples of estimated speed maps for typical weekday and weekend traffic in June 2010 are presented in the right-hand columns of Figure 27.

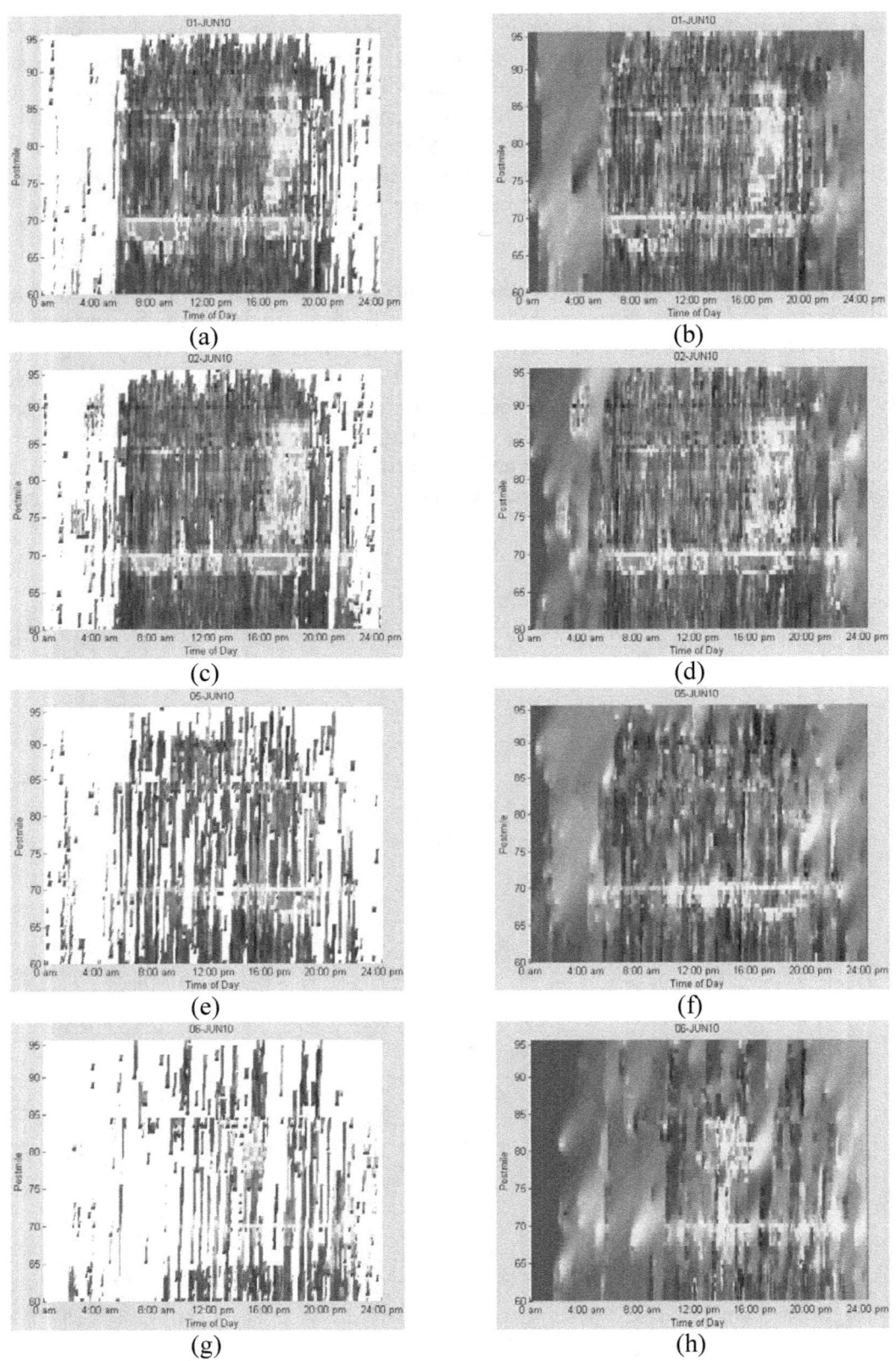

Figure 27. Samples of daily spatiotemporal speed map for Travel Database 2.

6.3 Test Environment

This section aims to investigate the performance of the proposed travel time prediction algorithm. In total, three case studies are conducted and the proposed algorithm is compared with different methods using the second travel database from 2010. The first travel database from October 2011 to November 2012 is expected to fuse with the 2010 data set to conduct more extensive tests in the future study. Since heavy traffic volume is usually observed along I-64 and I-264 heading to Virginia Beach during the summer season and on weekends, efficient and accurate travel time prediction can be helpful to travelers in planning their trips and reducing traffic congestion around the area. Considering that most of the congestion areas between Richmond and Virginia Beach are located before the Hampton Roads Bridge-Tunnel or along I-264, a 37-mile freeway stretch is selected as the test site for the first two case studies before the extensive testing on the entire 95-mile freeway. The 37-mile freeway stretch selected is from Newport News to Virginia Beach along I-64 and I-264 and includes 59 sections as shown in Figure 28. Eventually, the third case study is conducted on the entire 95-mile freeway from Richmond to Virginia Beach. The detailed information of experiment setups and discoveries are provided as below.

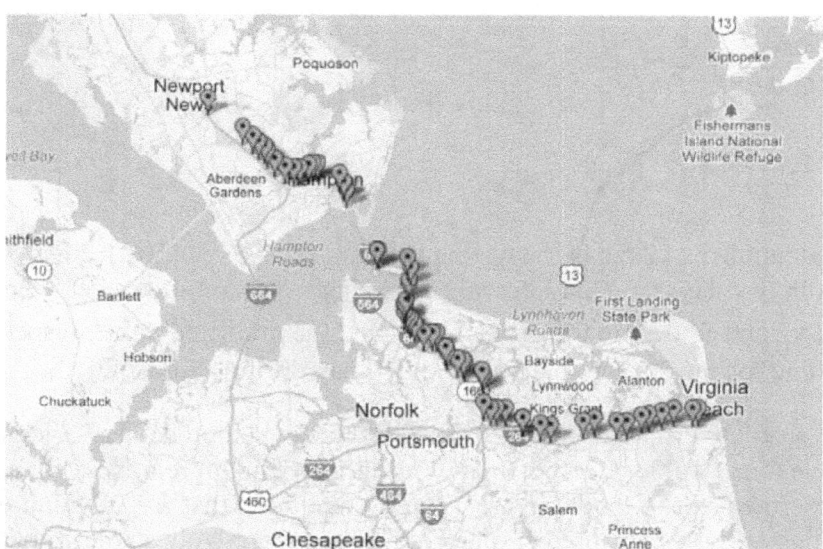

Figure 28. Selected 37-mile freeway stretch for case studies 1 and 2.

6.4 Case Study 1

Because traffic congestion for the selected freeway stretch is significant during the summer holiday season and on weekends, the evaluation of the travel-time prediction algorithm focuses on traffic data from June to August of 2010 in this case study. Traffic data from June and July are used for the training data set and the August data are used for the testing data set. The dynamic travel time is calculated every 5 minutes using the daily spatiotemporal traffic speed map, as shown in Figure 8, which serves as the ground truth data. The prediction span p equals zero for this test, which indicates that the future trip starting from the current time is the prediction output. The average travel time is predicted using Equation (173).

Different parameters are tested to identify the best combination to minimize the prediction error. The range of L, which represents the data length across the time axis (look ahead time duration), is between 10 to 60 minutes at 10-minute intervals. H is another parameter representing the shift distance across the time space when searching for a traffic map slice from the historical data set (look back duration). The size of H should not be too small. Otherwise, many overlapping map slices may be extracted for comparison to the current traffic map, and the computation time would be significant. Conversely, detailed information may be ignored if the value of H is too large. Therefore, the domain of the H value is also tested from 10 to 60 minutes at 10-minute increments.

Both relative and absolute prediction errors are calculated during this study to evaluate the proposed algorithm. The relative error is computed as the Mean Absolute Percentage Error (MAPE) using Equation (174). This error is the average absolute percentage change between the predicted and the true values. The corresponding absolute error is presented by the Mean Absolute Deviation (MAD) of Equation (175). This error is the absolute difference between the predicted and the true values.

$$MAPE = \frac{100}{I \times J} \sum_{j=1}^{J} \sum_{i=1}^{I} \frac{\left| y_i^j - \hat{y}_i^j \right|}{y_i^j}. \tag{174}$$

$$MAD = \frac{1}{J \times I} \sum_{j=1}^{J} \sum_{i=1}^{I} \left| y_i^j - \hat{y}_i^j \right|. \tag{175}$$

Here J is the total number of days in the testing the data set (i.e., 30 days); I is the total number of time intervals in one day (i.e., 204 intervals occurring every 5 minutes between 5:00 a.m. and 10:00 p.m.); and y_i^j and $\hat{y}_i^j$ denote the ground truth and the predicted value, respectively, of the dynamic travel time for the ith time interval on the jth day during August 2010.

The relative and absolute errors calculated by the proposed method across various parameters are presented in Table 6 and Table 7, respectively. Both the minimum relative error of 5.96% and the minimum absolute error of 2.96 min are obtained assuming that $L = 20$ minutes and $H = 40$ minutes. According to the tables, prediction errors are comparatively stable values of 6% and 3 min when L is less than 40 minutes. The change of the H value seems to have little impact on the average prediction accuracy under this situation. The optimum values of parameters can be used as a reference for applications on different sites.

Table 6. Relative errors by proposed travel time prediction method.

MAPE (%)		Time Interval of H (min.)					
		10	20	30	40	50	60
Time Interval of L (min)	10	6.09	5.98	6.13	5.98	6.00	6.03
	20	6.07	6.01	6.14	5.96	5.99	6.05
	30	6.17	6.05	6.14	5.99	5.97	5.98
	40	6.24	6.12	6.14	6.10	6.06	6.02
	50	6.27	6.15	6.20	6.15	6.21	6.12
	60	6.37	6.32	6.31	6.25	6.33	6.20

Table 7. Absolute errors by proposed travel time prediction method.

MAD (min.)		Time Interval of H (min.)					
		10	20	30	40	50	60
Time Interval of L (min)	10	3.02	2.98	3.05	2.99	2.99	3.00
	20	3.05	3.00	3.06	2.96	3.00	3.02
	30	3.11	3.04	3.08	3.01	3.00	3.00
	40	3.15	3.08	3.08	3.07	3.04	3.03
	50	3.17	3.09	3.11	3.10	3.12	3.09
	60	3.22	3.19	3.18	3.14	3.19	3.14

To better evaluate the proposed method in this case study, a traditional KNN algorithm (Qiao et al. 2012, Bustillos and Chiu 2011) is tested to predict travel time by applying the same training and testing data sets. However, instantaneous travel time is used in the KNN method instead of dynamic travel times as is used in the literature. Assuming the purpose is to predict the travel time starts from time interval t, the traditional KNN method uses the travel time sequence between recent past $t-L$ and time interval $t-1$ to find a similar data sequence in the historical data set. However, the dynamic travel time of the recent past travel time sequence may not be available since the trip has not been completed (the travel time is around 38 min under free-flow speed for the selected 37-mile freeway stretch). Therefore, instantaneous travel times between time interval $t-L$ and $t-1$ are used in the KNN method to predict travel time in the next time interval t.

Table 8. Relative errors by KNN method.

MAPE (%)		Time Interval of H (min.)					
		10	20	30	40	50	60
Time Interval of L (min)	10	6.80	6.68	6.85	6.68	6.70	6.74
	20	6.78	6.71	6.86	6.85	6.69	6.76
	30	6.61	6.59	6.61	6.69	6.66	6.68
	40	6.97	6.84	6.86	6.81	6.77	6.73
	50	7.01	6.87	6.93	6.87	6.94	6.83
	60	7.11	7.06	7.05	6.98	7.07	6.92

Table 9. Absolute errors by KNN method.

MAD (min.)		Time Interval of H (min.)					
		10	20	30	40	50	60
Time Interval of L (min)	10	3.51	3.53	3.55	3.51	3.53	3.52
	20	3.52	3.49	3.51	3.50	3.53	3.56
	30	3.52	3.48	3.47	3.51	3.56	3.54
	40	3.56	3.58	3.54	3.60	3.59	3.64
	50	3.59	3.61	3.58	3.67	3.64	3.68
	60	3.67	3.64	3.64	3.68	3.71	3.73

The same 10 closest candidates are selected using the Euler distance to calculate the average travel time for the future trip. The relative and absolute errors calculated by the proposed method across various parameters are presented in Table 8 and Table 9, respectively. The optimum parameter of L, which represents the domain of continuous time included in the traffic map slice, is 30 minutes; the corresponding minimum relative and absolute prediction errors are 6.59% and 3.47 min, respectively. Therefore, the average performance of the proposed method includes fewer errors compared to the traditional KNN method for our database. The main difference between the two methods is that the travel time sequence is used to obtain similar traffic patterns from historical data in the KNN method, while the traffic status across the spatial and temporal axes are used in the proposed method. The spatiotemporal traffic status provides more travel information given that it accounts for the spatial variation in the information. Therefore, such information serves a better pattern-matching result from the historical data and results in a more accurate travel time prediction performance. Moreover, the instantaneous travel time predicted by the KNN method may deviate substantially from the dynamic travel time under transient states during the trip. Based on the testing results, we observed that the predicted travel time using the KNN method is usually underestimated when congestion is forming and is overestimated when congestion is dissipating.

A comparison of the two methods for a typical weekday (i.e., August 2, 2010) is presented in Figure 29. The typical weekday traffic occurring on the selected 37-mile freeway stretch usually includes two peak hours during the morning and afternoon peaks. The traffic congestion is especially serious during afternoon peak hours. The ground truth curve in Figure 29 indicates that the travel time during this period could be more than two times (78 minutes) the travel time occurring during a free-flow period (38 minutes). The red curve obtained from the proposed method is a better fit to the ground truth data for congested and uncongested time periods. However, the blue curve obtained by the KNN method underestimates the actual travel time during congested afternoon periods and overestimates the actual travel time as the peak ends around 18:00 p.m. Consequently, the proposed method produces more accurate travel time prediction results compared to the KNN method for the subject day.

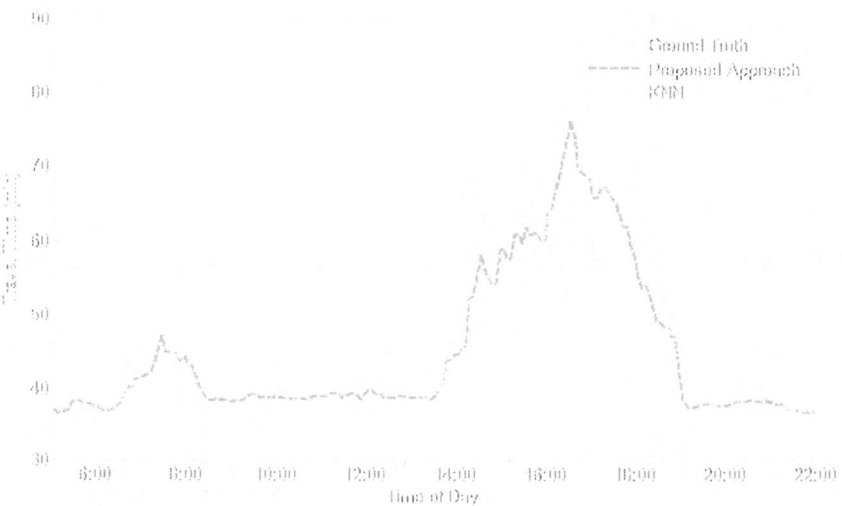

Figure 29. Comparison of prediction results for a typical weekday (August 2, 2010).

Another comparison of the two methods for typical weekend traffic occurring on August 7, 2010, is presented in Figure 30. Unlike typical weekday traffic, light traffic congestion occurs during the weekend that lasts for an extended time as many travelers go to Virginia Beach for that time period. Although the prediction accuracy is almost the same for this day when using the two methods, the green curve calculated by the traditional KNN approach also indicates that the deviation from ground truth data happens under transient states during which congestion is forming or dissipating.

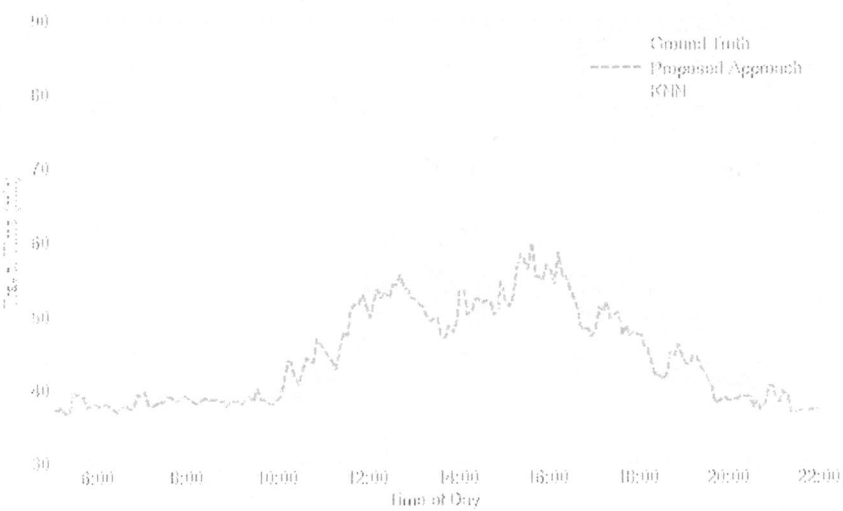

Figure 30. Comparison of prediction results for a typical weekend (August 7, 2010).

6.5 Case Study 2

Based on the results of case study 1, several improvements are adopted to improve the prediction accuracy of the proposed method. Consequently, a revised travel time prediction algorithm is proposed with the detailed revisions that have been described earlier. The selected 37-mile freeway stretch from Newport News to Virginia Beach is also used to investigate the performance of the revised prediction algorithm. Since serious congestion usually happens during the summer season on I-64 and I-264 heading to Virginia Beach, the traffic data on August 2010 is used as the testing data set. All the previous traffic data from April to July 2010 are used as the historical data set.

In order to better evaluate the performance of the proposed approach, three other methods from the previous studies are also tested on the same data set. Descriptions of all prediction methods are presented below.

1) Method 1: summation of the real-time traffic data across all the segments of freeway stretch.

2) Method 2: Kalman filter approach using the travel times from previous time intervals to define the transition function.

3) Method 3: k-nearest-neighbor method to predict the travel time by the selected similar travel time sequence from the historical data set.

4) Method 4: the proposed revised travel time prediction algorithm.

As with the previous case study, both absolute (mean absolute errors [MAE]) and relative (MAPE) errors are used to measure the prediction accuracy. The prediction results of the four methods are presented in Table 10. The prediction results are calculated for different days of a week (from Monday to Sunday), and the proposed method has the fewest prediction errors, except for Wednesday and Sunday. The average errors demonstrate that the proposed method is the best travel time prediction method. The best method for each row is highlighted by the blue color.

Table 10. Prediction results of four methods.

	Method1		Method2		Method3		Method4	
	MAE(min)	MAPE(%)	MAE(min)	MAPE(%)	MAE(min)	MAPE(%)	MAE(min)	MAPE(%)
Monday	2.07	4.57	2.37	5.25	1.99	4.44	1.68	3.75
Tuesday	2.14	4.86	2.47	5.60	2.17	4.91	2.02	4.48
Wednesday	2.70	5.77	3.06	6.53	2.63	5.64	2.70	5.76
Thursday	2.68	5.84	2.99	6.58	2.71	5.90	2.31	4.87
Friday	2.90	5.94	3.41	6.99	2.80	5.75	2.64	5.45
Saturday	2.64	5.69	3.23	6.94	2.61	5.62	2.45	5.22
Sunday	1.29	3.19	1.51	3.74	1.27	3.15	1.44	3.55
Average	2.34	5.12	2.72	5.95	2.31	5.06	2.18	4.73

Considering the real application, the accurate travel time prediction during congested time periods is more important to the traveler, since the traveler can change the trip schedule accordingly to avoid getting stuck in a traffic jam. On the other hand, the travel time prediction during uncongested time periods attracts less concern since it does not have much of an effect on the traveler's trip. In order to investigate the prediction results for congested and uncongested time periods, the absolute and relative errors are aggregated by every 2-hour interval, as presented in Table 11.

Table 11. Prediction results of four methods for different time periods.

		Method1		Method2		Method3		Method4	
		MAE(min)	MAPE(%)	MAE(min)	MAPE(%)	MAE(min)	MAPE(%)	MAE(min)	MAPE(%)
Monday	6 am - 8 am	1.63	3.77	1.95	4.55	1.58	3.61	1.91	4.42
	8 am - 10 am	1.10	2.73	1.42	3.54	1.15	2.86	0.91	2.28
	10 am - 12 pm	0.98	2.52	1.27	3.26	0.96	2.47	0.96	2.45
	12 pm - 14 pm	1.07	2.68	1.31	3.31	1.08	2.71	1.23	3.05
	14 pm - 16 pm	1.89	4.10	2.05	4.50	1.90	4.08	1.49	3.25
	16 pm - 18 pm	4.78	9.04	5.33	9.97	4.40	8.43	3.66	6.95
	18 pm - 20 pm	3.03	7.17	3.24	7.63	2.89	6.92	1.62	3.87
Tuesday	6 am - 8 am	1.78	4.15	2.16	5.07	1.78	4.13	1.64	3.89
	8 am - 10 am	1.32	3.36	1.57	3.97	1.41	3.56	1.18	2.97
	10 am - 12 pm	1.49	3.60	1.90	4.64	1.44	3.50	1.52	3.59
	12 pm - 14 pm	1.52	3.66	1.88	4.49	1.46	3.52	1.58	3.70
	14 pm - 16 pm	2.58	5.54	2.88	6.25	2.58	5.53	2.76	5.94
	16 pm - 18 pm	4.35	8.61	4.93	9.64	4.49	8.90	4.32	8.23
	18 pm - 20 pm	1.98	5.13	1.97	5.12	2.01	5.23	1.16	3.04
Wednesday	6 am - 8 am	1.46	3.57	1.81	4.44	1.41	3.46	1.51	3.75
	8 am - 10 am	2.07	4.60	2.19	4.96	2.06	4.59	2.20	4.90
	10 am - 12 pm	3.52	7.29	4.11	8.46	3.54	7.29	3.43	6.75
	12 pm - 14 pm	2.29	4.96	3.08	6.63	1.98	4.32	2.51	5.20
	14 pm - 16 pm	3.01	6.17	3.14	6.43	3.08	6.34	3.06	6.35
	16 pm - 18 pm	4.49	8.65	5.00	9.51	4.31	8.33	4.49	8.98
	18 pm - 20 pm	2.05	5.17	2.10	5.31	2.05	5.15	1.70	4.38
Thursday	6 am - 8 am	2.81	5.32	2.91	5.73	2.82	5.34	2.93	5.51
	8 am - 10 am	3.40	8.01	3.79	8.78	3.40	8.07	2.28	4.79
	10 am - 12 pm	1.11	2.82	1.56	3.96	1.14	2.88	1.07	2.68
	12 pm - 14 pm	1.42	3.53	1.60	3.97	1.44	3.59	1.70	4.17
	14 pm - 16 pm	2.42	5.33	2.65	5.99	2.50	5.49	2.71	5.90
	16 pm - 18 pm	4.44	8.27	5.02	9.34	4.60	8.56	3.57	6.48
	18 pm - 20 pm	3.14	7.62	3.37	8.26	3.03	7.38	1.88	4.56
Friday	6 am - 8 am	1.64	4.02	1.87	4.66	1.54	3.80	1.60	3.92
	8 am - 10 am	1.63	4.04	2.00	4.96	1.60	3.97	1.47	3.60
	10 am - 12 pm	1.57	3.76	1.85	4.43	1.56	3.74	1.83	4.38
	12 pm - 14 pm	2.66	5.75	3.18	6.89	2.66	5.79	3.40	7.13
	14 pm - 16 pm	4.57	8.54	5.32	9.89	4.18	7.84	3.85	6.96
	16 pm - 18 pm	4.61	7.78	5.73	9.80	4.35	7.30	4.21	6.86
	18 pm - 20 pm	3.58	7.71	3.89	8.30	3.67	7.85	3.23	6.67
Saturday	6 am - 8 am	1.03	2.72	1.14	3.00	1.00	2.63	0.72	1.90
	8 am - 10 am	0.90	2.36	1.14	2.96	0.94	2.44	0.88	2.28
	10 am - 12 pm	1.66	3.93	2.04	4.90	1.58	3.75	1.56	3.71
	12 pm - 14 pm	2.94	6.19	4.00	8.40	2.86	6.02	2.69	5.70
	14 pm - 16 pm	3.53	7.10	4.29	8.67	3.46	6.96	3.39	7.02
	16 pm - 18 pm	4.14	7.98	5.03	9.66	4.11	7.89	4.71	8.91
	18 pm - 20 pm	4.31	9.56	4.96	10.96	4.34	9.66	3.18	7.04
Sunday	6 am - 8 am	0.67	1.78	0.78	2.06	0.70	1.86	0.76	1.99
	8 am - 10 am	0.72	1.91	0.79	2.08	0.72	1.90	0.88	2.33
	10 am - 12 pm	0.94	2.44	1.11	2.89	0.92	2.41	0.81	2.11
	12 pm - 14 pm	1.06	2.66	1.34	3.38	1.02	2.56	1.12	2.74
	14 pm - 16 pm	1.65	3.98	1.91	4.64	1.55	3.77	2.12	5.11
	16 pm - 18 pm	1.29	3.25	1.59	4.03	1.25	3.17	1.59	4.06
	18 pm - 20 pm	2.69	6.29	3.05	7.08	2.74	6.40	2.83	6.50

According to the results in Table 11, the proposed method has the fewest prediction errors for most of the congested time periods, especially for evening peak hours (14 p.m. - 20 p.m.). For the uncongested time period, or the cases in which the proposed method does not work well, method 1 and method 3 produce the fewest errors. The detailed comparisons of prediction accuracy by different methods for every time period are demonstrated below.

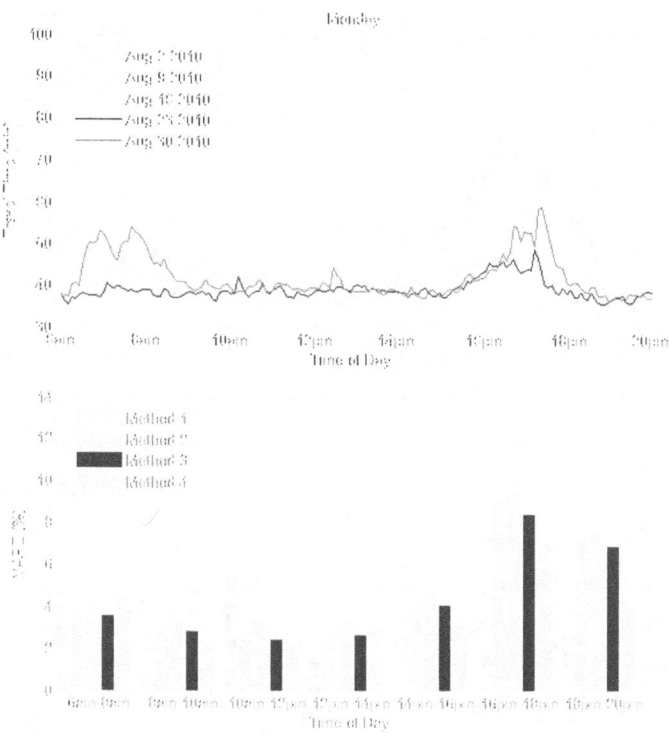

Figure 31. The ground truth travel time data and the comparison of four methods for Monday.

According to the travel time curves on Monday, there is mitigated congestion during morning peak hours (6 a.m. - 9 a.m.) and heavy congestion during afternoon peak hours (14 p.m. - 20 p.m.). The proposed method produces the fewest errors for the afternoon peak hours.

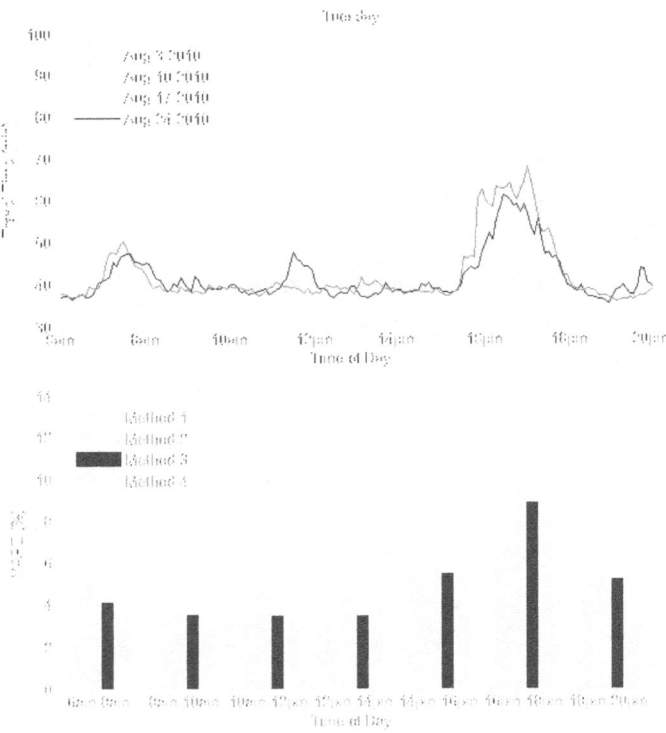

Figure 32. The ground truth travel time data and the comparison of four methods for Tuesday.

According to the travel time curves on Tuesday, there is mitigated congestion during morning peak hours (6 a.m. - 9 a.m.) and heavy congestion during afternoon peak hours (16 p.m. - 20 p.m.). The proposed method produces the least errors for both the morning and afternoon peak hours.

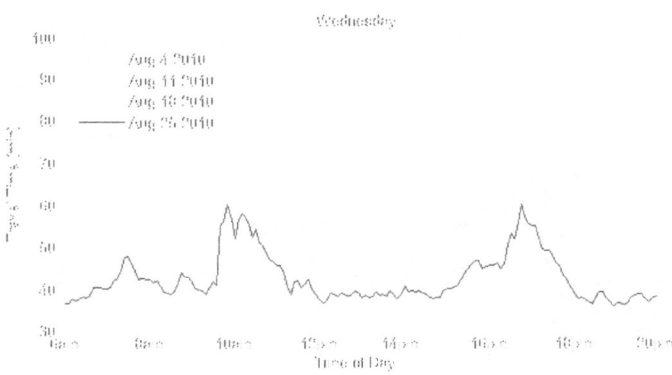

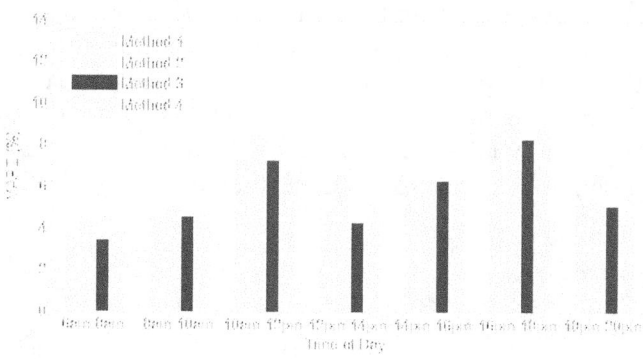

Figure 33. The ground truth travel time data and the comparison of four methods for Wednesday.

According to the travel time curves on Wednesday, there is mitigated congestion during morning peak hours (6 a.m. - 9 a.m.) and heavy congestion during afternoon peak hours (14 p.m. - 19 p.m.). The proposed method does not work well for both morning and afternoon peak hours, since there is abnormal congestion on August 18 (heavy congestion around 10a.m.) and August 25 (heavy congestion between 10 a.m. to 14 p.m.). Such abnormal congestion increases the difficulty of finding similar traffic patterns from the historical data set; therefore, the proposed method cannot produce the fewest prediction errors on Wednesday. This problem may be solved if such abnormal traffic data are included on our historical database.

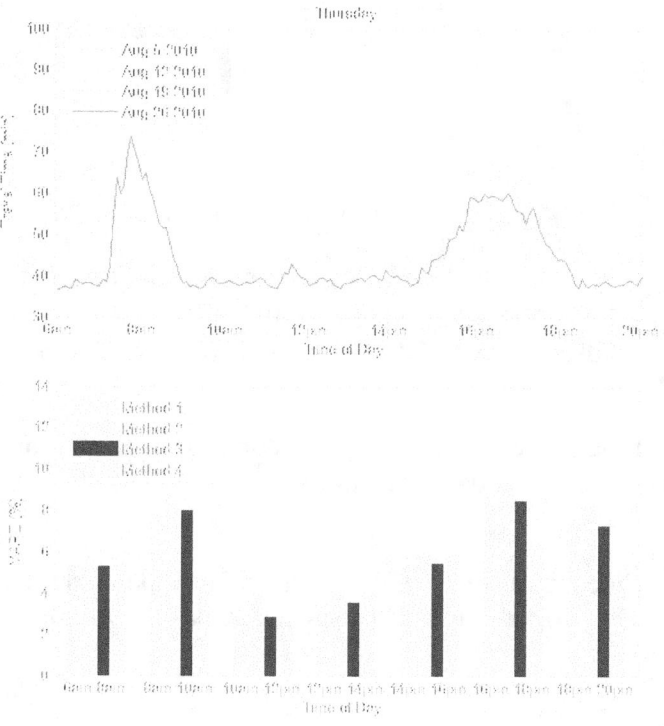

Figure 34. The ground truth travel time data and the comparison of four methods for Thursday.

According to the travel time curves on Thursday, there is mitigated congestion during morning peak hours (6 a.m. - 9 a.m.) and heavy congestion during afternoon peak hours (15 p.m. - 20 p.m.). The proposed method produces the fewest errors for both the morning and afternoon peak hours. The results from 6 a.m. - 8 a.m. and 14 p.m. - 16 p.m. demonstrate that the proposed method does not work very well when the congestion is forming; however, it works very well when congestion is dispatching, as compared to other methods.

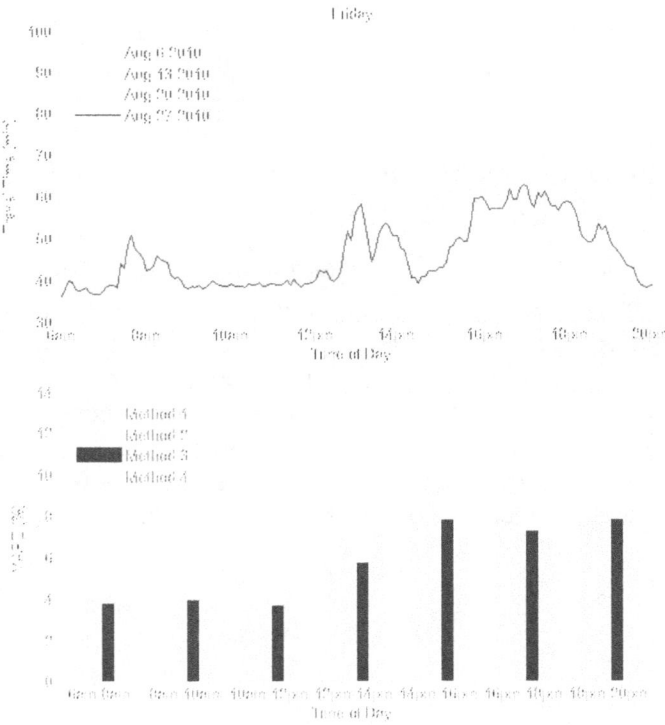

Figure 35. Method error relative to ground truth for Friday.

According to the travel time curves on Friday, there is mitigated congestion during morning peak hours (7 a.m. - 9 a.m.) and heavy congestion during afternoon peak hours (14 p.m. - 20 p.m.). The proposed method produces the fewest errors for both the latter half of the morning peak period and the afternoon peak hours.

According to the travel time curves on Saturday, the congestion occurs for a longer time from 11 a.m. to 20 p.m. than it does on weekdays. The proposed method works well, except the time period of 16 p.m. - 18 p.m.

According to the travel time curves on Sunday, congestion happens irregularly during the afternoon and evening. The proposed method does not work very well for such scenarios.

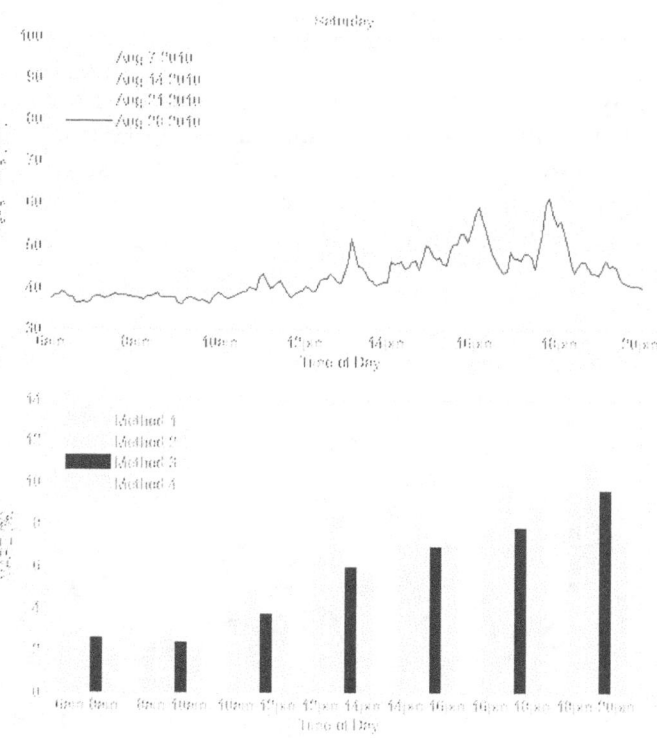

Figure 36. Method error relative to ground truth for Saturday.

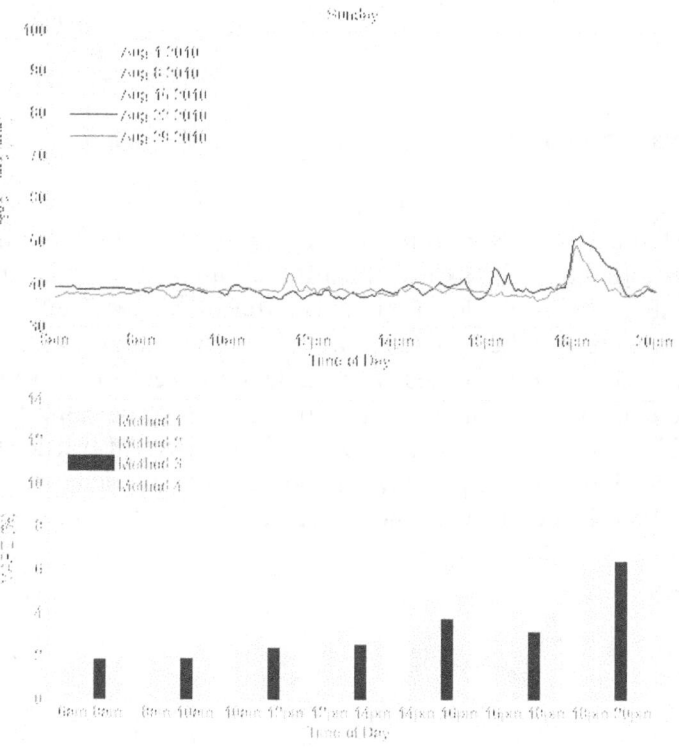

Figure 37. Method error relative to ground truth for Sunday.

6.6 Case Study 3

The entire 95-mile freeway stretch from Richmond to Virginia Beach is used in this case study to test the revised prediction algorithm. The same testing data set of 31 days in August 2010 as in the previous case studies is also used. All the previous traffic data since April 1, 2010 are used as the historical data set. For instance, if the current testing day is August 21, 2010, the historical data set is constituted by the 142 days from April 1, 2010 to August 20, 2010.

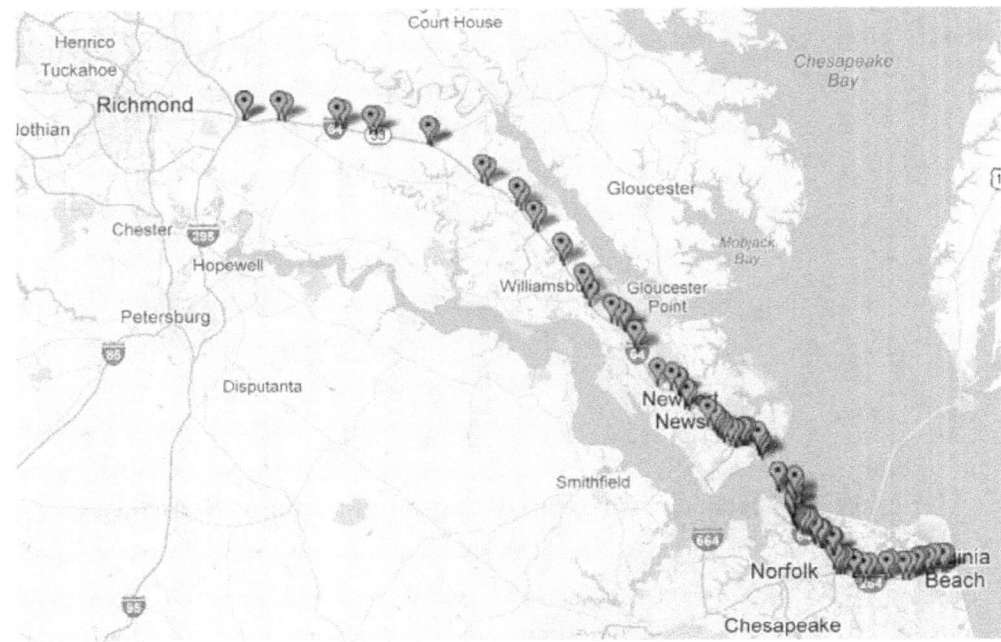

Figure 38. Freeway stretch from Richmond to Virginia Beach along I-64 and I-264.

In order to evaluate the performance of the proposed algorithm, the predicted travel time is compared with the instantaneous travel time as currently used by the Virginia Department of Transportation. For an actual trip, the experienced travel time is the actual travel time for a vehicle to travel from its origin to its destination. The roadway speed will not only change across space but also across time during the trip. Comparatively, the instantaneous travel time is the summation of real-time section travel times without the consideration of speed evolution across time. Many traffic agencies use instantaneous travel time to represent the future travel time information and provide this to drivers, which is valid if the current traffic status remains constant until the completion of the trip. However, instantaneous travel time may deviate substantially from the experienced travel time under transient states during which congestion is forming or dissipating.

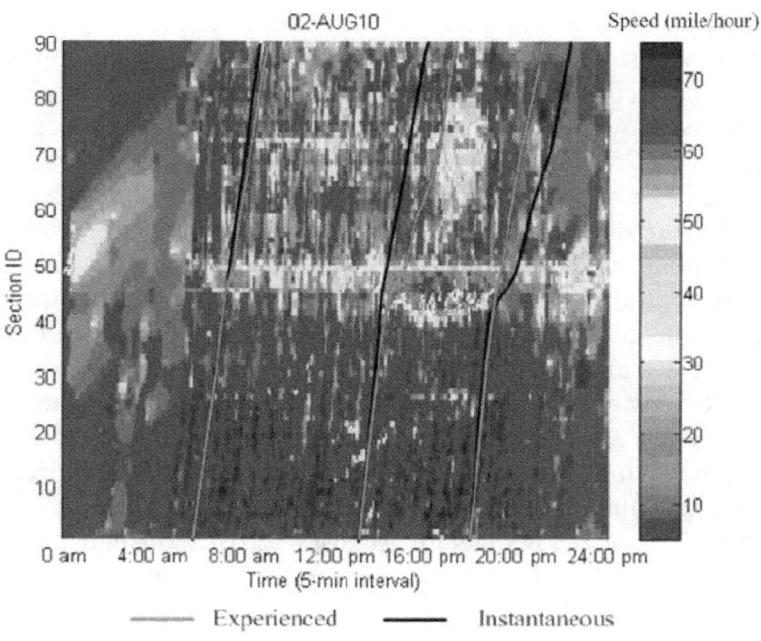

Figure 39. Spatiotemporal traffic state map and trip trajectories.

The temporal-spatial traffic state map is obtained from raw INRIX data on August 2, 2010 as presented in Figure 39. The differences between the instantaneous and experienced travel times are demonstrated by plotting the vehicle trajectories using both approaches. For a trip starting at 6:00 a.m., the instantaneous travel time is computed as 90 minutes (using the instantaneous speed values along all roadway segments at 6:00 a.m.) and the trip trajectory is depicted by the left black line. However, the corresponding experienced travel time (100 minutes) from the left red trajectory is 10 minutes longer than the instantaneous travel time, since the trajectory experiences the morning congestion (6:30 - 8:00 a.m.) before the tunnel along I-64. Large differences between instantaneous and experienced travel times occur during the afternoon peak hours because of heavy congestion. The instantaneous travel time starting at 14:00 p.m. is calculated as 92 minutes from the middle black trajectory, during which no congestion is included. But the experienced trajectory represented by the middle red curve is 117 minutes, since the trip encounters the tunnel congestion at around 14:30 p.m. and the low speed sections along I-264 at around 15:30 p.m. Therefore, the instantaneous travel time underestimates the experienced travel time by 25 minutes. The last demonstration is the opposite situation: the instantaneous travel time at 19:30 p.m. is 123 minutes, since the traffic is highly congested at two locations (tunnel and I-264). However, the experienced trajectory represented by the right red curve is 95 minutes and encounters almost no congestion. The instantaneous travel time overestimates the experienced travel time by 28 minutes in this case. To sum up, the above trajectories demonstrate that the instantaneous travel time calculated using the real-time data is not a good approach for predicting the experienced travel time, especially during congested periods.

The proposed approach is used to predict the temporal-spatial traffic states using historical data to construct vehicle trajectories and estimate travel times. The actual experienced travel time is calculated on the test data set as the ground truth. In order to effectively compare the instantaneous travel time results with the proposed approach, the congestion periods are extracted to calculate the prediction errors. The congestion status can be defined if the speed is less than 80% of the free flow speed. Therefore, the congested travel time can be identified if the travel time is higher than 1.25 times the free flow speed travel time. Generally, up to four congestion periods may be extracted during a day by aggregating the adjacent time intervals of congested status, which are morning, noon, afternoon, and evening congested periods. Considering the 31 days in August 2010, totally 57 congested periods are identified. The MAEs are calculated to compare the prediction accuracies between the predicted travel times and ground truth values during congested periods. By sorting the MAEs of instantaneous travel time and plotting the corresponding errors from two methods, the differences in prediction performance can be demonstrated in Figure 40. The average MAE by instantaneous travel time is 9.2 minutes and the average MAE from the proposed method is reduced to almost half error of 4.8 minutes. Among the 57 total congested periods, the proposed method produces many fewer errors (averaging less than 40%) than does the instantaneous travel time assumption for 49 periods. The samples of congested periods for August 4, 2010 and August 27, 2010 are highlighted in Figure 40. The proposed method outperforms instantaneous travel time during all congested periods for these two days.

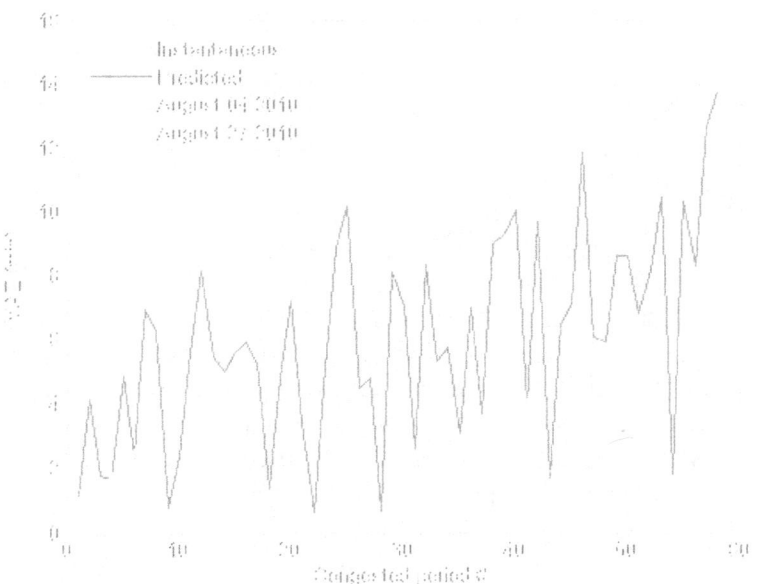

Figure 40. Average MAE of congested periods by two methods.

In order to investigate the maximum deviation between prediction results and ground truth data, the maximum MAE for using two methods for different testing days is selected and presented in Figure 41. The range of maximum MAE by instantaneous method is 11.7 - 44.5 minutes. The range of maximum MAE by the proposed method is 8.5 - 26.2 minutes, which is much lower than for the instantaneous method.

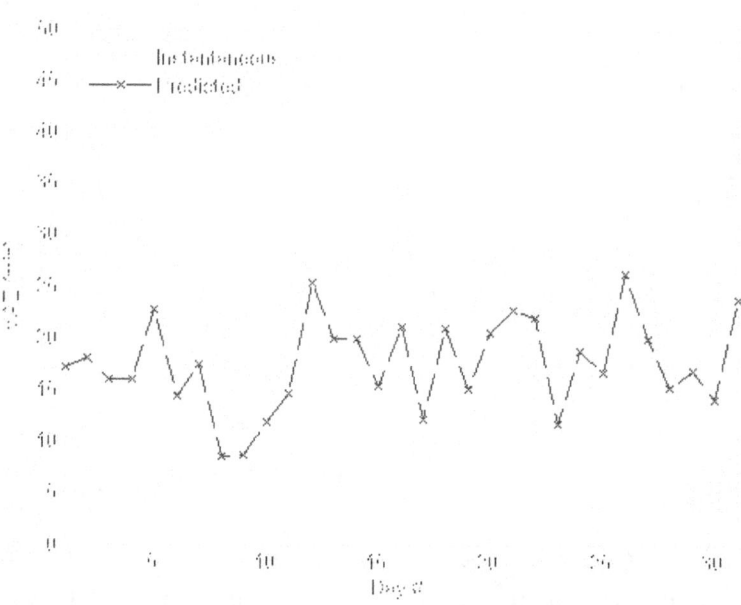

Figure 41. Maximum MAE by two methods for August 2010.

The travel time curves by the proposed prediction method, the instantaneous travel time, and the ground truth data for a typical weekday – August 4, 2010 –are presented in Figure 42 (a). The instantaneous travel time experiences some time lag to the ground truth data, especially during the time that congestion is forming or dissipating. Specifically, the instantaneous travel time highly underestimates the ground truth value when congestion is forming, and overestimates the ground truth travel time when congestion is dissipating. Comparatively, the proposed method improves the prediction performance when congestion is forming but still has some lag. The prediction performance during the congestion dissipating period is highly improved by the proposed method, since the propagation of shockwave can be predicted according to the historical trend from selected candidates. For instance, comparing the prediction errors by the proposed and instantaneous approach on August 4, 2010, the maximum reductions for congestion forming and dissipating periods are 12 minutes (from 13.4 to 1.4 minutes) at 14:30 p.m. and 25 minutes (from 25.8 to 0.8 minutes) at 17:40 p.m., respectively. The other benefit of the proposed approach is that travel time distribution can also be predicted other than as a deterministic value. The 95% and 5% confidence intervals of the predicted travel times are calculated as the upper and bottom boundaries as shown in Figure 42 (b). The green shadow area between boundaries covers most of the ground truth data curve, which demonstrates that the proposed approach provides very good accuracy to predict travel time reliability.

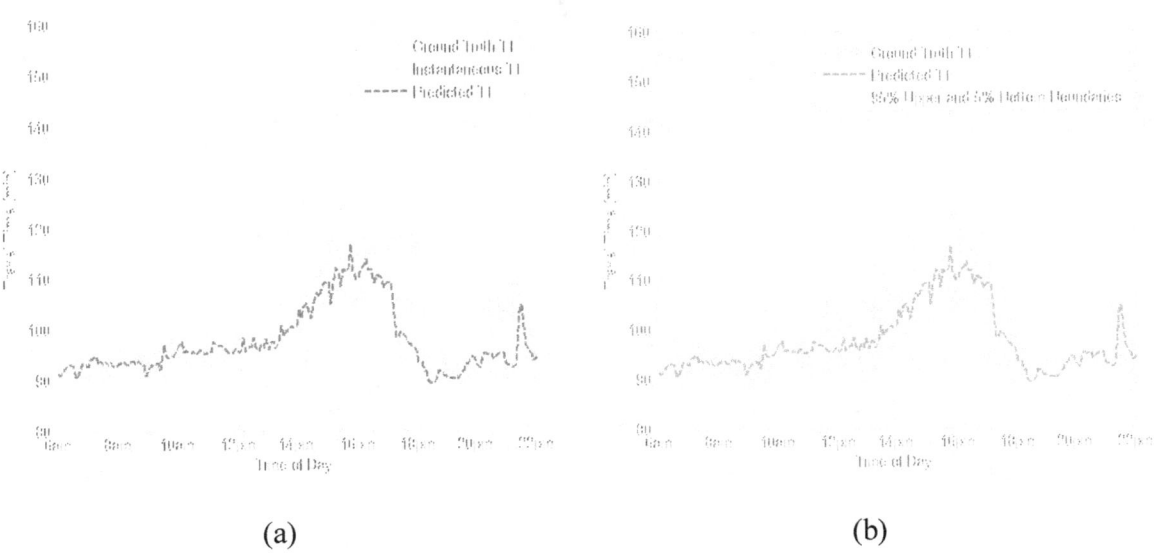

Figure 42. Travel time prediction results on August 04, 2010.
(a) Comparison between the proposed approach and instantaneous travel time; (b) The upper and bottom boundaries of proposed approach.

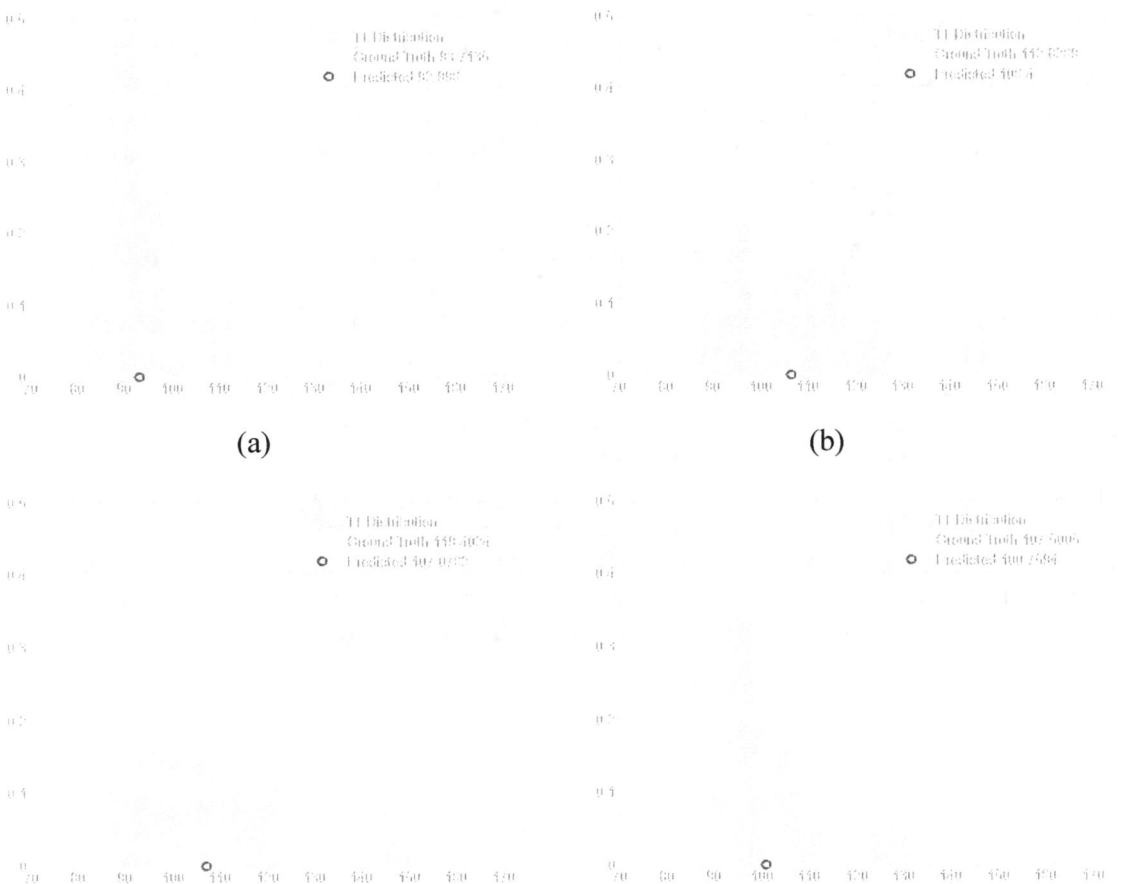

(c)   (d)

Figure 43. Predicted travel time distribution on August 04, 2010.
(a) 9 a.m. (b) 15 p.m. (c) 16 p.m. (d) 17 p.m.

Besides calculating the upper and bottom boundaries of travel time reliability, the travel time distribution can also be predicted by the proposed algorithm. Each selected candidate traffic status is corresponding to a predicted travel time and an associated weight value, which has been described in Equation (173). The weight value represents the dissimilarity between current and historical traffic patterns and can be used to calculate the probability of the associated travel time prediction result. Consequently, the predicted travel time distribution can be calculated for each time interval, as presented in Figure 43. The mean of travel time distribution is denoted by red dots and compared with the ground truth travel time value to demonstrate the high prediction accuracy. Rather than computing the mean value, other statistical representations can also be calculated as the prediction output. For instance, computing the 80th percentile in Figure 43 can be another option and maybe has less error to ground truth data. More importantly, the previous work of the research team to model travel time distribution can also be considered to improve prediction accuracy in the future study.

Similarly, the proposed approach outperforms the instantaneous approach on August 27, 2010, as shown in Figure 44. The maximum reductions of prediction errors by the proposed approach are 12.2 minutes (from 15 to 2.8 minutes) at 12:55 p.m. and 17.4 minutes (from 17.7 to 0.3 minutes) at 17:55 p.m. during congestion forming and dissipating periods, respectively.

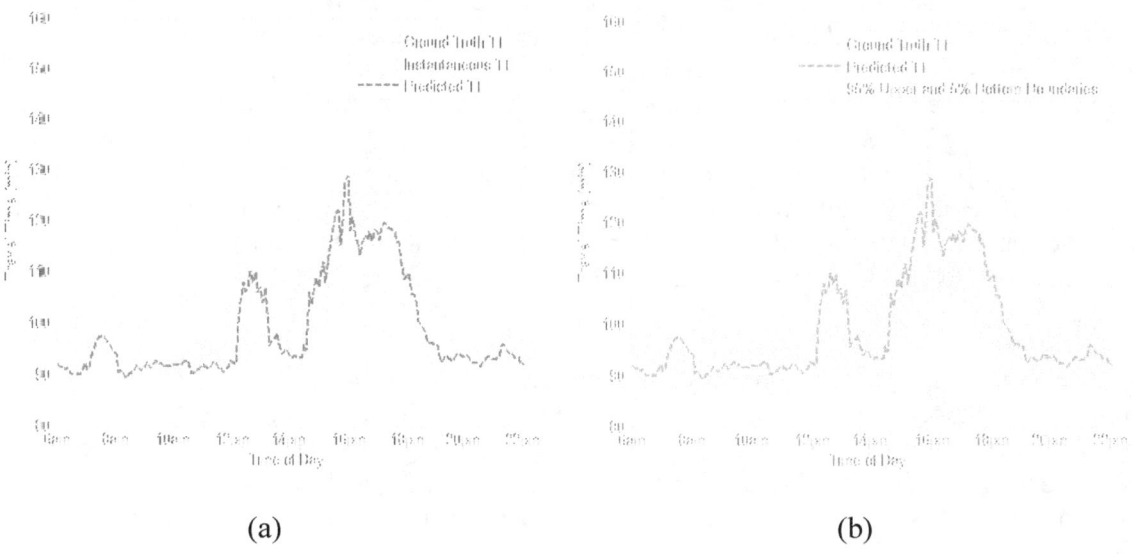

(a)   (b)

Figure 44. Travel time prediction results on August 27, 2010.
(a) Comparison between the proposed approach and instantaneous travel times; (b) The upper and lower boundaries of proposed approach.

7 Conclusions

7.1 Traffic Estimation

Based on the smoothness assumption and a kinematics principle a general first-order model relating directional derivatives of a priori travel time with speed at each point in the time-space domain is presented. This model due to its PDE form makes it possible to estimate travel times without the need for trajectory construction based on speeds which is in effect a cumbersome integration operation. A forward time backward space (FTBS) finite difference scheme is presented to approximate the general model.

While the general model and its corresponding FTBS solution scheme are designed for smooth situations, the real traffic stream is replete with discontinuities (shockwaves) especially as congestion increases. For these situations an equivalent conservative model is proposed. The model derivation is based on the first order continuum traffic flow model with Greenshields flux. A Godunov scheme is proposed to approximate the conservative model. This scheme is expected to provide a robust solution in presence of abrupt discontinuities in the solution domain.

In order to illustrate the performance of the proposed travel time models and their corresponding solution schemes, the US-101 dataset from NGSIM project is used. In essence, three factors affecting performance of the proposed schemes are investigated; that is speed accuracy, traffic congestion level, and discretization level.

To provide a benchmark for comparison, two sets of travel time estimates based on observed and estimated speeds are reported in the numerical experiments. These experiments showed that proposed schemes are indeed very accurate when accurate speeds are used. In fact, under this scenario travel time estimates with mean absolute percent error (MAPE) of about 6% and 10% were obtained using FTBS and Godunov schemes, respectively. However, travel time as an integral of traffic pace (inverse of speed) suffers from the presence of error in underlying speeds. Even though the mean absolute error (MAE) of speed estimates was in four to six mile per hour range, travel time estimates based on them were disproportionately less accurate. Under these circumstances FTBS and Godunov schemes have almost similar performance in terms of their accuracy.

In reported experiments congestion level and therefore the extent to which smoothness assumption is violated is varied in each of the three US-101 fifteen minute datasets. These datasets represent increasing congestion levels over time during a typical morning rush hour. Arguably, Godunov scheme exhibited a better performance at higher congestion levels as opposed to the FTBS scheme which quickly loses its edge as traffic conditions become more volatile.

Finally, specifics of the solution domain discretization which effectively determines the solution resolution is considered as an important factor which impacts the accuracy of proposed finite difference schemes. In the experiments two discretization stencils are examined. However, it should be noted that due to the limited length of the US-101 segment covered in the datasets as well as rather high free flow speed on this segment testing larger stencils was impossible. Despite these limitations, comparing errors at tested resolutions suggest that errors will decrease as the cell size increases. This observation is compatible with the fact that with increasing cell sizes variations inside the cells are smoothed out while discrepancies between adjacent cells become more pronounced. Therefore, it may be expected that at lower resolutions Godunov scheme will exhibit better performance compared to the FTBS scheme. However, this needs to be further verified in future experiments.

7.2 Dynamic Travel Time Prediction

The following conclusions can be drawn from the study of dynamic travel time prediction:

- Traffic state estimation and prediction form the basis of travel time prediction, given that spatiotemporal traffic state information is required to compute travel times.

- Dynamic travel times reflect experienced roadway travel times better than instantaneous travel times. This study is the first attempt to quantify and address such problems through the use of spatiotemporal traffic state information to predict dynamic travel times.

- Based on the reduction of INRIX data, a travel database was constructed for this study that includes daily spatiotemporal traffic state data (in this case, speed measurement) over the past 2.5 years along I-64 and I-264 (Richmond to Virginia Beach). The travel database is an important resource for future traffic applications in the Virginia Beach area.

- A new algorithm was developed that utilizes the spatiotemporal traffic-state information to predict travel times. Historical candidate data with similar traffic patterns to the current traffic status are selected using the proposed algorithm. These candidates are then aggregated to predict travel times.

- INRIX data for the selected 37-mile freeway stretch (Newport News to Virginia Beach) are used to test the prediction accuracy of the proposed algorithm. The results indicate that the proposed algorithm can accurately predict future travel times more accurately than three state-of-the-practice methods; those being: use of instantaneous measurements, a Kalman filter, and the k-nearest-neighbor method.

- The case study on the entire 95-mile freeway stretch from Richmond to Virginia Beach demonstrates the superiority of the proposed algorithm over the instantaneous approach, which is currently used by VDOT. Specifically, the proposed method reduces the prediction error by 50 percent compared to the state-of-the-practice instantaneous approach, especially at the shoulders of the peak periods.

- The proposed algorithm is flexible in terms of data resolution and sensing technology, and easily transferable to new locations. In addition, the proposed algorithm generates a travel time distribution as opposed to a single travel time estimate, which is typically done by current models.

The following are the recommendations resulting from the study of dynamic travel time prediction:

1. The proposed algorithm employed during this study provides a framework to use spatiotemporal traffic data to predict dynamic travel times. More advanced template matching or pattern recognition techniques should be considered and tested within the proposed algorithm to identify similar traffic patterns more efficiently and accurately.

2. The tests in this study only included summer data for a single year. The prediction accuracy is expected to improve if more summer data from previous years can be provided. In addition, different validation methods can be considered to better utilize the limited summer data, such as leave one out cross validation (LOOCV). In this approach, each summer day can be used as testing data while the remaining days are used as historical data.

3. The historical data set did not include weather, incident, or special day information. If such information can be used in the historical data set, the prediction of current day will be improved since a refined data set can be used to find similar traffic states more accurately. For instance, a subset of the database can be used for rainy conditions as opposed to using the entire data set. The development of such data will require the development of some clustering techniques to identify unique traffic state clusters.

4. Integrate loop detector data in addition to INRIX probe data. Traffic counts collected from loop detectors can be used to quantify ramp flows. Such information can be used with macroscopic traffic flow models to predict traffic states. In this way, both the advantages of macroscopic traffic modeling and the data-driven approach proposed in this study can be combined to predict travel times more accurately.

5. Develop algorithms to identify bottleneck locations using spatiotemporal speed data measurements to enhance the identification of similar traffic patterns. In addition, the impact of scheduled events on travel times should also be incorporated in the prediction algorithm, such as tunnel closures, etc. This can be done using the combined macroscopic traffic stream modeling and statistical modeling approaches.

6. Develop optimum methods to display travel time information. Since the proposed approach can be used in real-time applications and travel time distributions can be predicted instead of a single travel time estimate, the following aspects should be considered. Such as with the time interval to update predicted travel time, display average travel time, upper/lower travel time bounds, or 80th percentile travel times, etc.

7. Implement the proposed algorithm on different corridors to test the transferability of the approach. The predicted travel times on different corridors can be used to develop control strategies for route choice recommendations or area congestion reduction. A key input to these approaches is conducting research on how drivers respond to the provision of real-time information and how they switch their routes of travel depending on the information provided to them.

8. The problem of missing data should be examined during future studies considering that such problems are common in the field application. Different data estimation algorithms should be developed regarding various data sources and application requirements.

8 References

Ahmed, M. S., and A. R. Cook. "Analysis of Freeway Traffic Time Series Data by using Box-Jenkins Techniques." *Transportation Research Record* (Journal of Transportatin Research Board of the National Research Council) 722 (1979): 1-9.

Ahmed, S. A., and A. R. Cook. "Application of Time Series Analysis Techniques to Freeway Incident Detection." *Transportation Research Record* (Journal of Transportation Research Board of the National Research Council) 841 (1982): 19-21.

Astraita, A. "A Continuous Time Link Model for Dynamic Network Loading Based on Travel Time Functions." *Proceedings of the 13th International Symposium of Transportation and Traffic Theory.* Oxford, UK: Elsevier, 1996. 79-102.

Bajwa, S. I., E. Chung, and M. Kuwahara. "Performance Evaluation of an Adaptive Travel Time Prediction Model." *Proceedings of the 8th International IEEE Conference on Intelligent Transportation Systems.* Vienna, Austria, 2005.

Barcelo, J., L. Montero, L. Marques, and C. Carmona. "Travel Time Forecasting and Dynamic Origin Destinaiton Estimation for Freeways Based on Bluetooth Traffic Monitoring." *Transportation Research Record: Journal of the Transportation Research Board* (Transportation Research Board of the National Academies), no. 2175 (2010): 19-27.

Bardos, C., A. Y. Leroux, and J. C. Nedelec. "First Order Quasilinear Equations with Boundary Conditions." *Communications in Partial Differential Equations* 4, no. 9 (1979): 1017-1034.

Beskos, D. E., and P. G. Michalopoulos. "An Application of the Finite Element Method in Traffic Signal Analysis." *Mechanics Research Communications* 11, no. 3 (1984): 185-189.

Beskos, D. E., P. G. Michalopoulos, and J. K. Lin. "Analysis of Traffic Flow by the Finite Element Method." *Applied Mathematical Modeling* 9 (October 1985): 358-364.

Bickel, P. J., C. Chen, J. Kwon, J. Rice, J. Van Zwet, and P. Varaiya. "Measuring Traffic." *Statistical Science* 22, no. 4 (2007): 581-597.

Box, G. E.P., G. M. Jenkins, and G. C. Reinsel. *Time Series Analysis: Forecasting and Control.* 4th. Hoboken, New Jersey: John Wiley, 2008.

Cambridge Systematics Inc. "NGSIM U.S. 101 Data Analysis: Summary Report." Federal Highway Administration, Washington, D.C., 2005.

Cambridge Systematics, Texas Transportation Institute. *Traffic Congestion and Reliability: Linking Solutions to Problems.* Federal Highway Administration, U.S. Department of Transportation, Washington, D.C.: Office of Operations, 2004.

Carey, M. "Link Travel Times I: Desirable Properties." *Networks and Spatial Economics* 4, no. 1 (2004): 257-268.

Carey, M., and Y. E. Ge. "Retaining Desirable Properties in Discretising a Travel Time Model." *Transportation Research Part B* 41, no. 1 (2007): 540-553.

Carey, M., Y. E. Ge, and M. Mc Cartney. "A Whole Link Travel Time Model with Desirable Properties." *Transportation Science* 37, no. 1 (February 2003): 83-96.

Carmi, A., P. Gurfil, and D. Kanevsky. "Methods for Sparse Signal Recovery Using Kalman Filtering with Embedded Pseudo-Measurement Norms and Quasi-Norms." *IEEE Transactions on Signal Processing* 58, no. 4 (April 2010): 2405-2409.

Cassidy, M. J., and J. R. Windover. "Methodology for Assessing Dynamics of Freeway Traffic Flow." *Transportation Research Record* (Transportation Research Board of the National Research Council) 1484 (1995): 73-79.

Chakroborty, P., and S. Kikuchi. "Using Bus Travel Time Data to Estimate Travel Time on Urban Corridors." *Transportation Research Record: Journal of the Trasnportation Research Board* (Transportation Research Board of the National Research Council), no. 1870 (2004): 18-25.

Chen, M., and S. I.J. Chien. "Dynamic Freeway Travel Time Prediction with Probe Vehicle Data." *Transportation Research Record* (Journal of the Transportation Research Board of the National Research Council) 1768 (2001): 157-161.

Chu, L., J.-S. Oh, and W. Recker. "Adaptive Kalman Filter Based Freeway Travel Time Estimation." *Presented at Transportation Research Board Annual Meeting.* Washington, D.C.: Transportation Research Board of the National Academies, 2005.

Clark, S. "Traffic Prediction Using Multivariate Nonparametric Regression." *ASCE Journal of Transportation Engineering* 129, no. 2 (March-April 2003): 161-168.

Claudel, C. G., A. Hofleitner, N. D. Mignerey, and A. M. Bayen. "Guaranteed Bounds on Highway Travel Times Using Probe and Fixed Data." *Presented in Transportation Research Board Annual Meeting.* Washington, D.C.: Transportation Research Board of the National Academies, 2009.

Claudel, C. G., and A. M. Bayen. "Guaranteed Bounds for Traffic Flow Parameters Estimation Using Mixed Lagrangian-Eulerian Sensing." *46th Annual Allerton Conference on Communication, Control, and Computing.* Allerton, Illinois, 2008.

Claudel, C. G., and A. M. Bayen. "Lax-Hopf Based Incorporation of Internal Boundary Conditions into Hamilton-Jacobi Equation. Part I: Theory." *IEEE Transactions on Automatic Control* 55, no. 5 (May 2010): 1142-1157.

Claudel, C. G., and A. M. Bayen. "Lax-Hopf Based Incorporation of Internal Boundary Conditions into Hamilton-Jacobi Equation. Part II: Computational Methods." *IEEE Transactions on Automatic Control* 55, no. 5 (May 2010): 1158-1174.

Coifman, B. "Estimating Travel Times and Vehicle Trajectories on Freeways using Dual Loop Detectors." *Transportation Research Part A* 36, no. 4 (2002): 351-364.

Coifman, B. "Identifying the Onset of congestion Rapidly with Existing Traffic Detectors." *Transportation Research Part A* 37 (2003): 277-291.

Coifman, B., and E. Ergueta. "Improved Vehicle Reidentification and Travel Time Measurement on Congested Freeways." *ASCE Journal of Transportation Engineering* 129 (2003): 475-483.

Daganzo, C. "A Variational Formulation of Kinematic Waves Basic Theory and Complex Boundary Conditions." *Transportation Research Part B* 39 (2005): 187-196.

Daganzo, C. "A Variational Formulation of Kinematic Waves: Solution Methods." *Transportation Research Part B* 39 (2005): 934-950.

Daganzo, C. F. *Fundamentals of Transportation and Traffic Operations.* Oxford: Pergamon, 1997.

Daganzo, C. F. "The Cell Transmission Model: A Dynamic Representation of Highway Traffic Consistent with the Hydrodynamic Theory." *Transportation Research Part B* 28, no. 4 (1994): 269-287.

Dailey, D. J. "A Statistical Algorithm for Estimating Speed from Single Loop Volume and Occupancy Measurements." *Transportation Research Part B* 33, no. 5 (1999): 313-322.

Dailey, D. J. "Travel Time Estimation Using Cross Correlation Techniques." *Transportation Research Part B* 27B, no. 2 (1993): 97-107.

D'Angelo, M. P., H. M. Al-Deek, and M. C. Wang. "Travel Time Prediction for Freeway Corridors." *Transportation Research Record* (Journal of Transportation Research Board of the National Research Council) 1676 (1999): 184-191.

Dion, F., and H. Rakha. "Estimating Dynamic Roadway Travel Times Using Automatic Vehicle Identification Data for Low Sampling Rates." *Transportation Research Part B* 40 (2006): 745-766.

Fausett, L. V. *Fundamentals of Neural Networks: Architecture, Algorithms, and Applications.* Englewood Cliffs, New Jersey: Prentice-Hall, 1994.

Federal Highway Administration, U.S. Department of Transportation. *Next Generation SIMulation Fact Sheet.* December 2006. http://www.fhwa.dot.gov/publications/research/operations/its/06135/index.cfm (accessed May 10, 2011).

Greenshields, B. "A Study of Traffic Capacity." *Highway Research Board*, 1935: 448-477.

Haghani, A., M. Hamedi, and K. F. Sadabadi. *I-95 Corridor Coalition Vehicle Probe Project: Validation of INRIX Data July-September 2008.* Research Report, Civil Engineering Department, University of Maryland at College Park, I-95 Corridor Coalition, 2009.

Haghani, A., M. Hamedi, K. F. Sadabadi, S. E. Young, and P. Tarnoff. "Data Collection of Freeway Travel Time Ground Truth with Bluetooth Sensors." *Transportation Research Record* (Journal of the Transportation Research Board of the National Academies), no. 2160 (2010): 60-68.

Hall, F. L., and B. N. Persaud. "Evaluation of Speed Estimates Made with Single Detector Data from Freeway Traffic Management Systems." *Transportation Research Record* (Journal of Transportation Research Board of the National Research Council) 1232 (1989): 9-16.

Hamad, K., M. T. Shourijeh, E. Lee, and A. Faghri. "Near Term Travel Speed Prediction Utilizing Hilbert Huang Transform." *Computer Aided Civil and Infrastructure Engineering* 24 (2009): 551-576.

Handley, S., P. Langley, and F. A. Rauscher. "Learning to Predict the Duration of an Automobile Trip." *Proceedings of the 4th International Conference on Knowledge Discovery and Data Mining.* New York: AAAI Press, 1998.

Hazelton, M. L. "Estimating Vehicle Speed from Traffic Count and Occupancy Data." *Journal of Data Science* 2 (2004): 231-244.

Herrera, J. C., and A. M. Bayen. "Incorporation of Lagrangian Measurement in Freeway Traffic State Estimation." *Transportation Research Part B* 44 (2010): 460-481.

Hoffman, G., and J. Janko. "Travel Times as a Basic Part of the LISB Guidance Strategy." *Proceedings of the Third International Conference on Road Traffic Control.* London, England: Institution of Electrical Engineers, 1990.

Houston TranStar. *Houston TranStar and Bluetooth Monitoring.* 2011. http://traffic.houstontranstar.org/bluetooth/transtar_bluetooth.html (accessed May 1, 2011).

Ishak, S., and H. Al-Deek. "Performance Evaluation of Short Term Time Series Traffic Prediction Model." *ASCE Jouranl of Transportation Engineering* 128, no. 6 (November-December 2002): 490-498.

Julier, S. J., and J. J. LaViola. "On Kalman Filtering with Nonlinear Equality Constraints." *IEEE Transactions on Signal Processing* 55, no. 6 (June 2007): 2774-2784.

Kalman, R. "A New Approach to Linear Filtering and Prediction Problems." *ASME Journal of Basic Engineering* 82 (March 1960): 35-45.

Kalman, R., and R. Bucy. "New Results in Linear Filtering and Prediction Theory." *ASME Journal of Basic Engineering* 83 (March 1961): 95-108.

Krukjian, A., S. Gershwin, P. Houpt, A. Willsky, and E. Chow. "Estimation of Roadway Traffic Density on Freeways using Presence Detector Data." *Transportation Science* 14 (1980): 232-261.

Kwon, J., B. Coifman, and P. Bickel. "Day to Day Travel Time Trends and Travel Time Prediction from Loop Detector Data." *Transportation Research Record: Journal of the Transportation Research Board* (Transportation Research Board of the National Academies), no. 1717 (2000): 120-129.

LeVeque, R. J. *Numerical Methods for Conservation Laws.* Basel: Birkhauser Verlag, 1992.

Li, B. "A Non-Gaussian Kalman Filter with Application to the Estimation of Vehicular Speed." *Technometrics* 51, no. 2 (May 2009): 167-172.

Li, B. "On the Recursive Estimation of Vehicular Speed using Data from a Single Inductance Loop Detector: A Bayesian Approach." *Transportation Research Part B* 43 (2009): 391-402.

Lighthill, M. J., and G. B. Whitham. "On Kinematic Waves: II. A Theory of Traffic Flow on Long Crowded Roads." *Proceedings of the Royal Society of London*, 1955: 317-345.

Lindveld, C. D. R., R. Thijs, P. H. L. Bovy, and N. J. Van der Zijpp. "Evaluation of Online Travel Time Estimators and Predictors." *Transportation Research Record* (Journal of Transportation Research Board of the National Research Council) 1719 (2000): 45-53.

Liu, R.-X., H. Li, and Z.-F. Wang. "The Discontinuous Finite Element Method for Red-and-Green Light Models for the Traffic Flow." *Mathematics and Computers in Simulation* 56 (2001): 55-67.

Liu, Y., and G.-L. Chang. "Estimation of Freeway Travel Time Based on Sparsely Distributed Detectors." *Paper Presented at the 9th International Conference on Applications of Advanced Technologies in Transportation Engineering.* Chicago, 2006.

Liu, Y., P.-W. Lin, X. Lai, G.-L. Chang, and A. Marquess. "Developments and Applications of a Simulation Based Online Travel Time Prediction System: Traveling to Ocean City, Maryland." *Transportation Research Record: Journal of Transportation Research Board* (Transportation Research Board of the National Academies), no. 1959 (2006): 92-104.

Lu, Y., S.C. Wong, M. Zhang, and C.-W. Shu. "The Entropy Solutions for the Lighthill-Whitham-Richards Traffic Flow Model with a Dsicontinuous Flow-Density Relationship." *Transportation Science* 43, no. 4 (November 2009): 511-530.

Lu, Y., S.C. Wong, M. Zhang, C.-W. Shu, and W. Chen. "Explicit Construction of Entropy Solutions for the Lighthill-Whitham-Richards Traffic Flow Model with a Piecewise Quadratic Flow-Density Relationship." *Transportation Research Part B* 42 (2008): 355-372.

Mehran, B., M. Kuwahara, and F. Naznin. "Implementing Kinematic Wave Theory to Reconstruct Vehicle Trajectories from Fixed and Probe Sensor Data." *Procedia Social and Behavioral Sciences: 19th International Symposium on Transportation and Traffic Theory* 17 (2011): 247-268.

Mihaylova, L., and R. Boel. "A Particle Filter for Freeway Traffic Estimation." *43rd IEEE Conference on Decision and Control.* Atlantis, Paradise Islands, Bahamas, 2004.

Nam, D. H., and D. R. Drew. "Analyzing Freeway Traffic under Congestion: Traffic Dynamics Approach." *ASCE Journal of Transportation Engineering* ?, no. ? (May-June 1998): 208-212.

Nam, D. H., and D. R. Drew. "Automatic Measurement of Traffic Variables for Intelligent Transportation Systems Applications." *Transportation Research Part B: Methodological* 33, no. ? (1999): 437-457.

Nam, D. H., and D. R. Drew. "Traffic Dynamics: Method for Estimating Freeway Travel Times in Real Time from Flow Measurements." *ASCE Journal of Transportation Engineering* ?, no. ? (May-June 1996): 185-191.

Nanthawichit, C., T. Nakatsuji, and H. Suzuki. "Application of Probe Vehicle Data for Real Time Traffic State Estimation and Short Term Travel Time Prediction on a Freeway." *Transportation Research Record: Journal of Transportation Research Board* (Transportation Research Board of the National Academies), no. 1855 (2003): 49-59.

Newell, G. F. "A Simplified Theory of Kinematic Waves in Highway Traffic, Part I: General Theory." *Transportation Research* 27B (1993): 281-287.

Ni, D., and H. Wang. "Trajectory Reconstruction for Travel Time Estimation." *Journal of Intelligent Transportation Systems* 12, no. 3 (2008): 113-125.

Okutani, I., D. E. Beskos, and P. G. Michalopoulos. "Finite Element Analysis of Freeway Dynamics." *Engineering Analysis* 3, no. 2 (1986): 85-92.

Park, D., and L. R. Rilett. "Forecasting Multiple Period Freeway Link Travel Times Using Modular Neural Networks." *Transportation Research Record: Journal of Transportation Research Board* (Transportation Research Board of the National Research Council), no. 1617 (1998): 163-170.

Rao, S. S. *Applied Numerical Methods for Engineers and Scientists.* Upper Saddle River, New Jersey: Prentice Hall, 2002.

Research and Innovative Technology Administration. *Pocket Guide to Transportation.* U.S. Department of Transportation. 2010. http://www.bts.gov/publications/pocket_guide_to_transportation/2010/ (accessed April 25, 2011).

Richards, P. I. "Shock Waves on the Highway." *Operations Research* 4 (1956): 42-51.

Rilett, L. R., and D. Park. "Direct Forecasting of Freeway Corridor Travel Times Using Spectral Basis Neural Networks." *Transportation Research Record: Journal of Transportation Research Board* (Transportatin Research Board of the National Research Council), no. 1752 (2001): 140-147.

Robinson, S., and J. W. Polak. "Modeling Urban Link Travel Time with Inductive Loop Detector Data by Using the k-NN Method." *Transportation Research Record: Journal of the Transportation Research Board* (Transportation Research Board of the National Academies), no. 1935 (2005): 47-56.

Sadabadi, K. F., and A. Haghani. "Real Time Solution of Velocity Based First Order Continuum Traffic Model Using Finite Element Method." *Transportation Research Record: Journal of the Transportation Research Board* (Transportation Research Board of the National Academies) xxxx (2011): xx-xx.

Sethian, J. A. "Fast Marching Methods." *SIAM Review* 41, no. 2 (June 1999): 199-235.

Simon, D. *Optimal State Estimation: Kalman, H[infinity] and Nonlinear Approaches.* Hoboken, New Jersey: John Wiley & Sons, 2006.

Strikwerda, J. C. *Finite Difference Schemes and Partial Differential Equations*. Belmont, CA: Wadsworth & Brooks/Cole, 1989.

Sun, L., J. Yang, and H. Mahmassani. "Travel Time Estimation Based on Piecewise Truncated Quadratic Speed Trajectory." *Transportation Research Part A* 42, no. ? (2008): 173-186.

Sun, X., L. Munoz, and R. Horowitz. "Mixture Kalman Filter Based Highway Congestion Model and Vehicle Density Estimator and its Application." *Proceedings of the American Control Conference*. Boston, Massachusetts, 2004.

Transportation Network Modeling Committee. *A Primer for Dynamic Traffic Assignment*. Washington, D.C.: Transportation Research Board of the National Academies, 2010.

Treiber, M., and D. Helbing. "Reconstructing the Spatio Temporal Traffic Dynamics from Stationary Detector Data." *Cooper@tive Tr@nsport@tion Dyn@mics* 1, no. 3 (2002): 1-24.

Tsitsiklis, J. N. "Efficient Algorithms for Globally Optimal Trajectories." *IEEE Transactions on Automatic Control* 40, no. 9 (september 1995): 1528-1538.

Van Hinsbergen, C. P.IJ., and J. W.C. Van Lint. "Bayesian Combination of Travel Time Prediction Models." *Transportation Research Record: Journal of Transportation Research Board* (Transportation Research Board of the National Academies), no. 2064 (2008): 73-80.

Van Trier, J., and W. W. Symes. "Upwind Finite Difference Calculation of Travel Times." *Geophysics* 56, no. 6 (June 1991): 812-821.

Vanajakshi, L., and L. R. Rilett. "System Wide Data Quality Control of Inductance Loop Data Using Nonlinear Optimization." *ASCE Journal of Computing in Civil Engineering* 20, no. 3 (May-June 2006): 187-196.

Vanajakshi, L., B. M. Williams, and L. R. Rilett. "Improved Flow Based Travel Time Estimation Method from Point Detector Data for Freeways." *ASCE Journal of Transportation Engineering* 135, no. 1 (January 2009): 26-36.

Waller, S. T., Y. -C. Chiu, N. Ruiz-Juri, A. Unnikrishnan, and B. Bustillos. *Short Term Travel Time Prediction on Freeways in Conjunction with Detector Coverage Analysis*. University of Texas, Austin: Texas Department of Transportation and the Federal Highway Administration, 2007.

Wang, Y., and M. Papageorgiou. "Real Time Freeway Traffic State Estimation Based on Extended Kalman Filter: A General Approach." *Transportation Research Part B* 39 (2005): 141-167.

Wang, Y., M. Papageorgiou, and A. Messmer. "Real Time Freeway Traffic State Estimation Based on Extended Kalman Filter: A Case Study." *Transportation Science* 41, no. 2 (2007): 167-181.

Wong, G.C.K., and S.C. Wong. "A Wavelet-Galerkin Method for the Kinematic Wave Model of Traffic Flow." *Communications in Numerical Methods in Engineering* 16 (2000): 121-131.

Work, D. B., O.-P. Tossavainen, S. Blandin, A. M. Bayen, T. Iwuchukwu, and K. Tracton. "An Ensemble Kalman Filtering Approach to Highway Traffic Estimation Using GPS Enabled Mobile Devices." *Proceedings of the 47th IEEE Conference on Decision and Control*. Cancun, Mexico, 2008. 5062-5068.

Ye, Z. R., Y. L. Zhang, and D. R. Middleton. "Unscented Kalman Filter Method for Speed Estimation using Single Loop Detector Data." *Transportation Research Record* (Journal of Transportation Research Board of the National Research Council) 1968 (2006): 117-125.

You, J., and T. J. Kim. "Development and Evaluation of a Hybrid Travel Time Forecasting Model." *Transportation Research Part C* 8 (2000): 231-256.

Yu, J., G.-L. Chang, H.W. Ho, and Y. Liu. "Variation Based Online Travel Time Prediction Using Clustered Neural Networks." *Paper Presented at the 11th International IEEE Conference on Intelligent Transportation System.* Beijing, China, 2008.

Zhang, X., and J. A. Rice. "Short Term Travel Time Prediction." *Transportation Research Part C* 11 (2003): 187-210.

Zou, N., J. Wang, G.-L. Chang, and J. Paracha. "A Hybrid Model for Reliable Travel Time Estimation on a Freeway with Sparsely Distributed Detectors." *Paper Presented at the ITS World Congress.* Beijing, China, 2007.

Zou, N., J. Wang, G.-L. Chang, and J. Paracha. "Application of Advanced Traffic Information Systems: Field Test of a Travel Time Prediction System with Widely Spaced Detectors." *Transportation Research Record: Journal of Transportation Research Board* (Transportation Research Board of the National Academies), no. 2129 (2009): 62-72.

Bustillos, B. I. & Y.-C. Chiu (2011) Real-Time Freeway-Experienced Travel Time Prediction Using N-Curve and k Nearest Neighbor Methods. *Transportation Research Record: Journal of the Transportation Research Board,* 2243, 127-137.

Chen, H. & H. A. Rakha. 2012. Prediction of Dynamic Freeway Travel Times based on Vehicle Trajectory Construction. In *15th International IEEE Conference on Intelligent Transportation Systems.*

Chen, H., H. A. Rakha & S. A. Sadek. 2011. Real-time Freeway Traffic State Prediction: A Particle Filter Approach. In *14th International IEEE Conference on Intelligent Transportation Systems,* 626-631. Washington, DC, USA.

Chen, H., H. A. Rakha, S. A. Sadek & B. J. Katz. 2012. A Particle Filter Approach for Real-time Freeway Traffic State Prediction. In *91st Transportation Research Board Annual Meeting.* Washington D.C.

Cheng, P., Z. Qiu & B. Ran. 2006. Traffic Estimation Based on Particle Filtering with Stochastic State Reconstruction Using Mobile Network Data. In *Transportation Research Board Annual Meeting.*

Chu, J. 2011. Travel Time Messages on Dynamic Message Signs.

Du, L., S. Peeta & Y. H. Kim (2012) An Adaptive Information Fusion Model to Predict the Short-term Link Travel Time Distribution in Dynamic Traffic Networks. *Transportation Research Part B: Methodological,* 46, 235-252.

Fei, X., C.-C. Lu & K. Liu (2011) A Bayesian Dynamic Linear Model Approach for Real-time Short-term Freeway Travel Time Prediction. *Transportation Research Part C: Emerging Technologies,* 19, 1306-1318.

Hinsbergen, C. P. I. J. v., A. Hegyi, J. W. C. v. Lint & H. J. v. Zuylen (2011) Bayesian Neural Networks for the Prediction of Stochastic Travel Times in Urban Networks. *IET Intelligent Transport Systems,* 5, 259-265.

Lint, J. W. C. v., S. P. Hoogendoorn & H. J. v. Zuylen (2005) Accurate Freeway Travel Time Prediction with State-space Neural Networks Under Missing Data. *Transportation Research Part C: Emerging Technologies,* 13, 347-369.

Mazare, P.-E., O.-P. Tossavainen, A. Bayen & D. Work. 2012. Trade-offs between Inductive Loops and GPS Vehicles for Travel Time Estimation: A Mobile Century Case Study. In *Transportation Research Board 91st Annual Meeting*. Washington, D.C.

Mihaylova, L., R. Boel & A. Hegyi (2007) Freeway Traffic Estimation within Particle Filtering Framework. *Automatica,* 43, 290-300.

Myung, J., D.-K. Kim, S.-Y. Kho & C.-H. Park (2011) Travel Time Prediction Using k Nearest Neighbor Method with Combined Data from Vehicle Detector System and Automatic Toll Collection System. *Transportation Research Record: Journal of the Transportation Research Board,* 2256, 51-59.

Nanthawichit, C., T. Nakatsuji & H. Suzuki (2003) Application of Probe-vehicle Data for Real-time Traffic-state Estimation and Short-term Travel-time Prediction on a Freeway. *Transportation Research Record: Journal of the Transportation Research Board*, 49-59.

Peeta, S. & J. J. L. Ramos. 2006. Driver response to variable message signs-based traffic information. In *IEEE Proceedings of Intelligent Transport Systems*, 2-10.

Qiao, W., A. Haghani & M. Hamedi. 2012. Short Term Travel Time Prediction Considering the Weather Impact. In *Transportation Research Board 91st Annual Meeting*. Washington D.C.

Rakha, H. & B. Crowther (2002) Comparison of Greenshields, Pipes, and Van Aerde Car-Following and Traffic Stream Models. *Transportation Research Record,* 1802.

Ristic, B., and Sanjeev Arulampalam. 2004. *Beyond the Kalman Filter: Particle Filters for Tracking Applications*. Boston, MA.

Sau, J., N.-E. E. Faouzi, A. B. Aissa & O. d. Mouzon. 2007. *Particle Filter-Based Real-Time Estimation and Prediction of Traffic Conditions*. Chania, Greece.

Schrank, D. & T. Lomax. 2007. 2007 Urban Mobility Report. Texas Transportation Institute

Tu, H. 2008. Monitoring Travel Time Reliability on Freeways. In *Department of Transport and Planning*. Technische Universiteit Delft.

Vanajakshi, L. & L. R. Rilett. 2007. Support Vector Machine Technique for the Short Term Prediction of Travel Time. In *IEEE Intelligent Vehicles Symposium*, 600-605. Turkey.

Vlahogianni, E. I., J. C. Golias & M. G. Karlaftis (2004) Short-term Traffic Forecasting: Overview of Objectives and Methods. *Transport Reviews,* 24, 533-557.

Wang, Y. & M. Papageorgiou (2005) Real-Time Freeway Traffic State Estimation Based on Extended Kalman Filter: A General Approach. *Transportation Research Part B,* 39, 141-167.

Wang, Y., M. Papageorgiou & A. Messmer (2008) Real-Time Freeway Traffic State Estimation Based on Extended Kalman Filter: Adaptive Capabilities and Real Data Testing. *Transportation Research Part A,* 42, 1340-1358.

Work, D. B., S. Blandin, O.-P. Tossavainen, B. Piccoli & A. M. Bayen (2010) A Traffic Model for Velocity Data Assimilation. *Applied Mathematics Research Express,* 1-35.

Work., D. B., O.-P. Tossavainen, S. Blandin, A. M. Bayen, T. Iwuchukuw & K. Tracton. 2008. An ensemble Kalman filtering approach to highway traffic estimation using GPS enabled mobile devices. In *47th IEEE Conference on Decision and Control*, 5062.

Wu, C.-H., J.-M. Ho & D. T. Lee (2004) Travel-time Prediction with Support Vector Regression. *IEEE Transactions on Intelligent Transportation Systems,* 5, 276-281.

Xia, J. & M. Chen. 2009. Dynamic Freeway Corridor Travel Time Prediction Using Single Inductive Loop Detector Data. In *Transportation Research Board 88th Annual Meeting*. Washington D.C.

Xia, J., M. Chen & W. Huang (2011) A Multistep Corridor Travel-Time Prediction Method Using Presence-Type Vehicle Detector Data. *Journal of Intelligent Transportation Systems: Technology, Planning, and Operations,* 15, 104-113.

Yang, J.-S. 2005. Travel Time Prediction Using the GPS Test Vehicle and Kalman Filtering Techniques. In *Proceedings of the 2005 American Control Conference*, 2128-2133.

Yang, M., Y. Liu & Z. You (2010) The Reliability of Travel Time Forecasting. *IEEE Trans. Intell. Transport. Syst.,* 11, 162-171.

www.ingramcontent.com/pod-product-compliance
Lightning Source LLC
Chambersburg PA
CBHW080301180526
45167CB00006B/2616